A. Ravishankar Rao

A Taxonomy for Texture Description and Identification

With 79 Illustrations

Springer-Verlag
New York Berlin Heidelberg
London Paris Tokyo Hong Kong

A. Ravishankar Rao
IBM T.J. Watson Research Center
Yorktown Heights, NY 10598
USA

Series Editor

Ramesh C. Jain
Electrical Engineering and
  Computer Science Department
University of Michigan
Ann Arbor, MI 48109
USA

Library of Congress Cataloging-in-Publication Data
Ravishankar Rao, A.
   A taxonomy for texture description and identification / A.
Ravishankar Rao.
      p.   cm. — (Springer series in perception engineering)
   Includes bibliographical references.
   1. Computer vision.   2. Visual texture recognition.   I. Title.
II. Series.
TA1632.R28   1990
006.3'7—dc20   90-35180

Printed on acid-free paper.

Photocomposed copy prepared using the author's LaTeX file.
Printed and bound by Edwards Brothers, Inc., Ann Arbor, Michigan.
Printed in the United States of America.

9 8 7 6 5 4 3 2 1

ISBN 0-387-97302-8 Springer-Verlag New York Berlin Heidelberg
ISBN 3-540-97302-8 Springer-Verlag Berlin Heidelberg New York

Dedicated to Amma, Appa, and Madan for all the support and
encouragement I have received.

Series Preface

Texture plays a very important role in many disparate machine perception tasks. Inspection of many products including lumber, semiconductor wafers, and textiles is facilitated by texture. Texture is also important in recovering surface depth information. Many endeavors in computer vision have been aimed at the classification of texture. Most of these efforts have been based on the statistical or frequency domain characteristics of images. Recently, however, some efforts have been directed towards characterizing oriented texture patterns in images.

Ravi Rao addresses the problem of texture classification and description in his book. He proposes a taxonomy for texture and has applied his classification system to several applications, including semiconductor wafer inspection. He also proposes approaches to describe complex compositions of textures.

Rao's work is strongly motivated by the desire to develop a language to describe visual patterns that can be used by people using machine vision systems. A visual language will facilitate interaction between users and the system and will result in user-friendly machine vision systems. The effort to develop expert systems in several image-based domains have clearly demonstrated the importance of such a language. Though this book addresses only texture, it is a good beginning in the direction of defining such a language.

The most important contribution of Rao's work is his novel method for a symbolic description of oriented texture patterns. He proposed characterizing oriented patterns using the qualitative differential equation theory. His algorithm starts with images, recovers dominant directions of intensity values in images, and then partitions the image using standard symbols derived from the theory. The parameters for these symbols are estimated. An image is thus partitioned into regions that are described by appropriate descriptors. This approach is an important step in bridging the signal-and-symbol gap. The efficacy of this approach is demonstrated by the consideration of several examples drawn from diverse applications. This approach to describing oriented patterns is not only related to oriented texture patterns, but will contribute significantly to applications such as the analysis of optical flow patterns and fluid flow visualization.

This work demonstrates how powerful concepts from established branches of mathematics can be used in pragmatic machine perception tasks. Considering the fact that most machine perception algorithms based on rigorous mathematical approaches are applicable only in very limited cases (mostly synthetic images) the performance of this algorithm is particularly impressive. Yet, the fact that it works so well is not surprising. The robustness of this algorithm is due to its emphasis on using rigorous mathematical tools to identify stable patterns and then using established techniques to estimate the parameters of those patterns in images.

I believe that the techniques presented in this book will be applicable in many areas. In machine perception, these techniques will soon find numerous applications in inspection and dynamic vision.

Ramesh Jain
Ann Arbor, Michigan

Preface

The invention of the camera and the microscope have resulted in a vast amount of visual imagery which is increasingly encountered in several disciplines ranging from medicine to semiconductor manufacturing. The advent of the computer introduces the challenge of trying to create and interpret these visual images in a meaningful manner.

Though these important tools are available to handle visual data, there is still no systematic framework to render an abstraction of such data. This problem can be compared to the evolution of language as a means for communication and expression. Language as we know it is primarily verbal. What is now needed is a *visual* language, one which is capable of representing visual data. Donis Dondis, in her book entitled *A Primer for Visual Literacy*, has made a strong effort to educate all people to their maximum ability, both as makers and receivers of visual messages. The underlying philosophy of this book has been to inculcate some of this spirit, and investigate the possibility of imparting visual literacy to computer vision systems.

There are very practical reasons for why this is a significant problem. Machine vision systems are becoming increasingly popular because of the high level of flexibility they add to automation. This trend will inevitably result in extensive and sophisticated interactions of vision systems with humans and other automated systems. A concrete example of such interactions is that of interfacing machine vision with diagnostic expert systems to perform automated process control in semiconductor wafer manufacturing. In fact, my involvement with this problem started through an interdisciplinary program initiated by the Semiconductor Research Corporation at the University of Michigan, whose goal is the development of techniques for the automation of semiconductor fabrication processes. The idea of designing a visual language was born out of the problems faced when trying to precisely characterize different kinds of surface defects that arise in semiconductor wafer processing.

In order to better understand the problem of designing a visual language, this book focuses on an important element of vision, namely that of texture. Texture occurs in a wide variety of visual input, and is also an important diagnostic in the manufacturing world. However, humans are rather poor at characterizing texture in a precise manner, be it qualitative or quantitative. The first step to tackle this problem is to devise a symbolic representation that is capable of capturing the qualitative aspects of texture. The next step is to develop computational techniques that will allow computers to automatically derive the relevant representation.

The book follows this approach towards understanding texture. The emphasis has been on finding the appropriate mathematical models for different kinds of texture, and developing computational techniques based on these models in order to automatically create a symbolic description for texture.

Acknowledgements

This book is the product of interactions with several people from different disciplines, to whom I am very grateful. This not only made the problem very interesting but also provided the impetus to complete the research work presented here.

At first, I would like to express my deep gratitude towards Prof. Ramesh Jain, for constant encouragement through all the stages that this book went through. I greatly appreciate his openness towards problems and their solutions, and for being patient and supportive during the formative stages. He gave me the freedom to develop these ideas, and provided the guidance needed to make it a reality.

I am grateful to Prof. Brian Schunck for many interesting discussions. Some of these led to a paper, which forms an important chapter in this book. I thank Prof. Dahm for his keen sense of understanding of some of the problems I posed. His insight led to the symbolic description scheme that I have used in this book for oriented textures. I wish to thank Prof. Weymouth and Prof. Wise who were on my doctoral committee, for their advice and encouragement.

I am grateful to Prof. Romesh Saigal for providing me the suggestion of using linear phase portraits to analyze oriented textures, and to Prof. Ben Van der Plujm, and Prof. Maria Comninou for discussions during the early phases.

I wish to acknowledge the continued support of the Semiconductor Research Corporation throughout my five years as a graduate student. This not only made the research possible, but also provided a valuable source of practical problems and considerations. Al Sharp from Texas Instruments helped me appreciate the complexity of the problem of automation in inspection. Wendy Fong and Bob Conder from Hewlett Packard brought the problem of defect description and identification to my attention. I am grateful to them for the same.

I thank Ali Kayaalp and Chuck Cole for helpful discussions. I thank Mike Barnes and Karen Studer-Rabeler and H. Chau for providing help in obtaining SEM images. I would also like to acknowledge useful input from several members of the Center for Integrated Circuits and Sensors at the University of Michigan.

I would like to thank my colleagues from the Artificial Intelligence Laboratory, Amir Amini, Arnold Chiu, Shih-Ping Liou, Sarvajit Sinha, Saeid Tehrani and Arun Tirumalai for several discussions and suggestions at various stages of this book. Paddy, Dilip and Michaelangelo provided useful comments, for which I am thankful. I am grateful to Kathy, Sara, Dolores and others of the ATL building for their help and goodwill, which contributes to the congenial environment that exists here.

I greatly appreciate the virtuosity of Vilayat Khan, whose music alleviated a considerable amount of the stress associated with research and writing.

I am glad to have met Rajeev, who proved to be a genuine friend through the ups and downs of graduate student life. Roukoz, Jim, LC and others from the EECS building made going to work a pleasant and enjoyable event.

I thank anonymous reviewers from the computer vision community for their comments on portions of this book which were sent out as papers to journals and conferences.

I wish to acknowledge the use of the facilities at the I.B.M. T. J. Watson Research Center while completing this book.

I am particularly grateful to Nalini Bhushan for several philosophical discussions about my work, as well as for moral support during the difficult period of the writing of this book.

My brother, Madan has helped shape my attitude towards work and all its associated problems. I much appreciate his zest for life in all its complexity. Finally, I wish to thank my parents for their wisdom and guidance at every stage in my life. Without their support and encouragement, none of this would have been possible.

Contents

"The illiterate of the future will be ignorant of pen
and camera alike."

Laszlo Moholy-Nagy (renowned Bauhaus artist), 1935

coarse

striate

speckled

wrinkled

granulated

rough

mottled

striped

smooth

worm-hole

speedboat

embayed

starburst

dentate

lamellar

orange peel

List of Figures

List of Tables

1

Introduction

One of the central issues in computer vision is the problem of signal to symbol transformation [44, 102], which really has two components. The first is the choice of symbols to be derived. This is usually ignored or taken for granted. The second is the exact method used to transform the given image into this set of symbols. Researchers in computer vision have usually focussed on this part of the problem. In this book we analyze the symbolic description of textures by first proposing an appropriate set of symbols and then solving the signal to symbol transformation problem for a specific type of texture called oriented texture.

Both the set of symbols and transformation techniques vary widely, depending on the visual cue that is the subject of analysis. For instance, if we are interested in 3D shape description, then the eight different surface types based on surface curvature sign, viz. *peak, pit, ridge, valley, saddle-ridge, saddle-valley, flat* and *minimal* [14] would constitute the symbol system. The idea behind such a representation is that arbitrary smooth surfaces may be decomposed into a union of simple surface primitives, as shown in the results of segmentation [14]. Another choice of symbols arises if we assume that the surface is quadric [15].

Similarly, color is characterized in terms of brightness, hue and saturation of the RGB components [128]. Due to this three dimensional nature of color, traditional color specification systems use a triple of numbers to describe color (see [48, 34] for a discussion of conversions between various systems).

Unfortunately, such a parallel does not exist for texture – we do not know what the dimensions of texture are. For texture, the choice of a symbol set is itself a difficult task. Though various researchers have realized the need for a description scheme for textures, they have *only* speculated that it should be organized hierarchically [46, 114]. As a result, there is no satisfactory symbol set to describe different kinds of textures. There is the additional problem of finding a relationship between symbolic descriptors for texture and computational techniques to abstract such descriptions. As Haralick [46] remarks, "... Unfortunately, few experiments have been done attempting to map semantic meaning into precise properties of tonal primitives and their spatial distributional properties...".

1.1   Scope of the book

There are three broad categories of texture problems [73, p. 424]. The first involves the identification and description of two-dimensional patterns, and has been widely addressed [46, 125]. Figure 1.1 summarizes the different approaches to texture description, as reviewed in [46].[1]

The second is concerned with using texture as a means to perform segmentation of an image [12, 121, 97, 79]. The third problem is to use texture as a cue to retrieve information about surface orientation and depth [2, 4, 3, 67, 127]. The problem I address in this book falls in the first category.

1.1.1   DEFINING TEXTURE

There are several branches of engineering and natural sciences that make extensive use of textural cues in identifying and describing different objects [23, 139, 78, 110, 25, 123]. Texture is also widely used in art and design [47, 53]. However, texture means something different to each area. Consequently, the word *texture* has a rich range of meanings, and it is worthwhile to examine some of them.

In mechanical engineering , *surface texture* is the repetitive or random deviation from the nominal surface that forms the 3D topography [139]. . In petrology, textures are the geometrical relationships among the component crystals of a rock and any amorphous materials that may be present (size of the component crystals, shape, distribution and orientation) [110, 78]. In art and design, the predominant dimension attached to texture is that of 'surface quality related closely to our sense of touch', [53], or 'the tactile quality of the surface', [87].

Webster's dictionary defines texture to be 'the visual or tactile surface characteristics and appearance of something'. This calls for a distinction between the **visual** and **tactile** dimensions of texture. As an illustration of this point, consider the texture of a reptile skin. Apart from being smooth (the tactile dimension) it has also got cellular and specular markings on it that form a well defined pattern (the visual dimension).

In this book, we will restrict ourselves to the *visual* dimension of texture, as commonly done in computer vision. Interestingly, most researchers in computer vision do not attempt to define texture in their work. Interpretations of what constitutes texture vary widely. According to Haralick, texture is characterized by tonal primitive properties as well as spatial relationships between them [46]. Horn [52] considers texture to be 'detailed structure in an image that is too fine to be resolved, yet coarse enough

---

[1]The review of texture research presented in this book is not intended to be comprehensive.

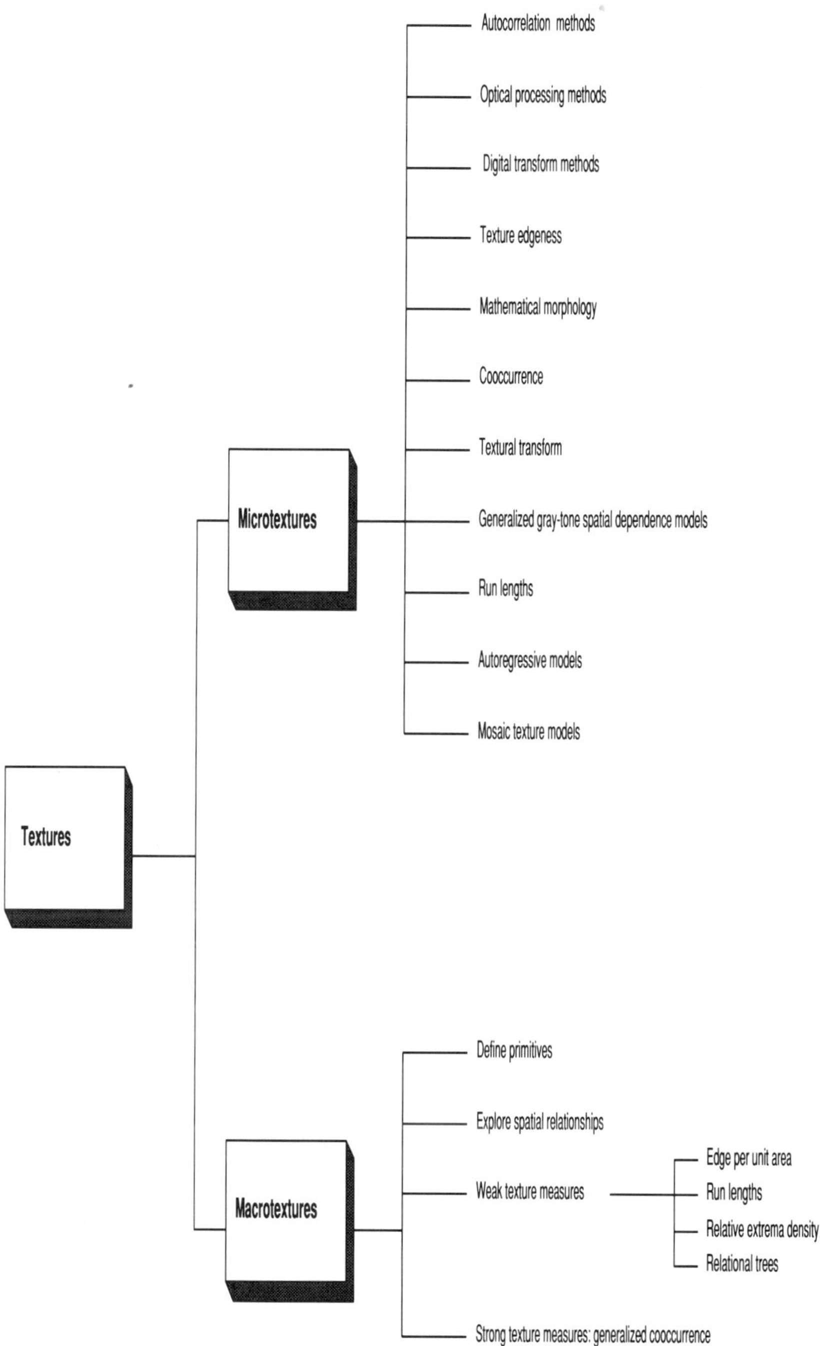

FIGURE 1.1. Different approaches to texture description

to produce a noticeable fluctuation in the gray levels of neighboring cells.' Some researchers [77] assume the dominant feature of texture to be the repetition of a pattern.

We adopt a more general view of texture, and consider it to be 'the surface markings or 2D appearance of a surface.' A constructive definition of texture will be given in chapter 6.

1.1.2   PROBLEM DEFINITION

The problem addressed in this book is to devise a symbolic representation for texture, and define computational paradigms to automatically derive this symbolic representation.

Thus, there are two aspects to the problem – qualitative and quantitative. The qualitative aspect is concerned with the design of a semantically relevant symbolic representation for texture. The quantitative aspect is concerned with measures for texture.

Though there is a rich body of literature on the human perception of textural features [61], my attempt in this book to describe texture is based purely on engineering grounds. The main reason for this is that studies of human texture perception have relied primarily on the *pre-attentive* features (Julesz and Bergen [61] define preattentive texture perception to occur when textures are viewed in a quick glance with little effort or analysis). Furthermore, these studies only require the discrimination of different textures, *without necessitating any sort of symbolic description*. In contrast to this, the inspection of textured surfaces, arising in many situations in manufacturing requires a considerable amount of scrutiny and effort. In such cases, it is useful for texture analysis to ultimately result in a symbolic description, or interpretation [104]. Hence I have searched for appropriate mathematical models that will enable one to render symbolic descriptions of given textures.

1.2   Importance of texture

Texture is an important element of vision and has been analyzed in its own right for the last two decades by researchers in psychophysics as well as computer vision [10, 11, 46, 61].

There are several disciplines and applications in which texture plays an important role. Texture plays a critical role in inspecting surfaces that are produced at various stages in all types of manufacturing [23]. For instance, in the inspection of semiconductor devices, surface texture is an important factor that is used to decide the integrity of a fabricated device [88]. Texture plays an important role in the area of machined parts. In fact, an entire standard [139] is devoted to the specifications related to surface texture. When a metal is deformed, its grains are reoriented, with certain orientations being preferred over other. This development of a preferred orientation gives rise to a texture. The analysis of such directional textures is important because many material properties, such as tensile strength, depend on the distribution of grain directionality [71, p. 14]. There are several areas like petrography [78, 110], metallography [123] and lumber processing [23] that make extensive use of textural features such as grain shapes, sizes, and distributions for recognizing and analyzing specimens. Texture is very important in quality control since many inspection decisions are based on the appearance of the texture of some material [35, 103] Subsurface microstructure, consisting of different kinds of textures, plays a crucial role in the analysis of fractograph specimens [137, pp. 27–35]. Special textures arise when using the method of flow visualization, which is becoming increasingly popular in diverse areas such as biomedical engineering, tracer methods, oceanography, aerodynamics, and surface flows [129].

1.3   Potential applications of this research

There are several possible applications of the ideas contained in this book. We present a cross-section of these applications, ranging from the general to the specific.

A long term goal of the research in this book is to form the basis of a visual language by illustrating the formation of a description scheme for texture. This motivation arose by comparing the concepts of verbal and visual literacy as follows. In order to attain *verbal* literacy, it is necessary to learn the basic components of a written language, such as the letters, words, spelling, grammar, syntax. Once this is mastered, a wide range of concepts may be expressed. *Visual* literacy is similar to verbal literacy. According to Dondis [31], "the purpose of *visual* literacy is to construct a basic system for learning, recognizing, making and understanding visual messages that are negotiable by all people, not just those specially trained, like the designer, the artist, the craftsman, and the aesthetician." Thus, the design of a visual language would enable both humans and computers to interpret different kinds of visual input in a meaningful manner.

As a step towards creating a visual language, this book presents a taxonomy for the description of texture (chapter 6). This taxonomy provides standardized symbolic descriptors that can be used for descriptions of a wide variety of textures. For instance, the taxonomy can be used as a basis to classify different defects arising in the fabrication of semiconductor devices. Preliminary results for this application are described in [103]. This will be useful in making a systematic catalog of defect types for diagnostic purposes. The taxonomy could also be used to cut down operator training time for defect inspection.

The automation of lumber defect detection is vital for the future control of lumber processing, which is an important industry [113]. Defects in wood are rich in textural content, and typically involve the identification of knots, worm holes, checks, and grains [25, 96, 33, 23]. The method that we present in chapter 3 is able to identify the locations of knots in pieces of wood. Furthermore, a precise characterization of each knot is available, which enables the program to reconstruct the original knot.

Flow visualization techniques are becoming increasingly popular in diverse areas such as biomedical engineering, tracer methods, oceanography, aerodynamics, and surface flows [129]. The analysis of flow-like patterns can form the basis of a scheme for flow visualization [57, 89]. Computer vision techniques can extract qualitative information about flow behaviour which may not be readily available from conventional flow measurements [129]. Image processing techniques have been applied to quantitative flow analysis with promising results [55], and there is still plenty of room for improvements in these techniques.

However, the full use of the quantitative information present in a flow visualization picture presents a difficult problem [50]. In order to tackle this problem, Imaichi and Ohmi [56] describe a method for the numerical processing of flow visualization pictures. The marriage of the scheme we have proposed with Imaichi's and Ohmi's work could result in a novel technique for the automated analysis of flow visualization pictures.

Though the algorithm presented in chapter 3 has been developed for the 2D case, it can be easily extended to the 3D case. This would enable the algorithm to also be used on 3D data that may be generated in flow visualization.

Another application to which the method in chapter 3 can be applied is fingerprint identification and recognition [101]. Since a fingerprint appears like a phase portrait for spiral flow, one can use the quantitative measures associated with this flow as features for the fingerprint.

The methods proposed in this book could also be applied to analyze subsurface microstructure. These microstructures consist of largely flow-like textures and plays a crucial role in the analysis of fractograph specimens [137]. The scheme presented in chapter 3 to describe these textures would be useful for the purposes of identification.

The discipline of petrography deals with several kinds of rock textures [78]. Texture is used to classify different types of rocks. For instance, orientation analysis is helpful in labeling certain textures as lamellar (plate-like) [78]. This could be done using the techniques developed in chapter 2.

The algorithms developed in this book could also be applied to the areas mentioned in section 1.2.

1.4   Issues in automated process control involving computer vision

Computer vision is being used to automate several kinds of manufacturing problems, such as semiconductor wafer inspection [30], and lumber processing [23]. In the domain of automated inspection, a crucial concern is the identification and description of defects. A large number of defects that occur and need to be identified are rich in textural information, and this is the case across different domains [23, 123, 40]. For instance, in the inspection of semiconductor devices, surface texture is an important factor that is used to decide the integrity of a fabricated device [88].

The impetus for this book was derived from a very practically motivated application of computer vision to manufacturing. We are conducting research to automate semiconductor wafer inspection by using machine vision [66, 59]. This is part of a larger venture, whose goal is to develop techniques for the automation of semiconductor fabrication processes.

1.4.1   BACKGROUND

In the semiconductor IC manufacturing facilities of the future, individual fabrication processes will be controlled by intelligent systems. These systems will receive information from sensors and systems for automatic inspection of wafers for determining the state of a process. In addition to quality control, inspection systems will be developed to monitor different aspects of wafers during manufacturing by performing critical measurements. Thus, automatic inspection systems will act as an intelligent front end, providing early feedback from the manufacturing process to a diagnostic expert system. The expert system embodies knowledge about the processing parameters to prevent mass production of bad chips.

The Hewlett Packard Photolithography Advisor [35] is an example of the usage of diagnostic expert systems for process control in semiconductor manufacturing. It is interesting to note that visual input to the expert system is provided manually. The system prompts the user for descriptions of different kinds of defects, and also has images of typical defects for the sake of comparison. However, the automation of this task of defect description and identification remains a challenging problem. The important issue here is how information should be exchanged between the visual inspection system and the diagnostic system.

1.4.2   PROBLEMS INVOLVING TEXTURE

Texture plays a critical role in inspecting surfaces that are produced at various stages in the inspection of semiconductor devices. For instance, surface roughness is a measure of the correctness of a reactive ion etch

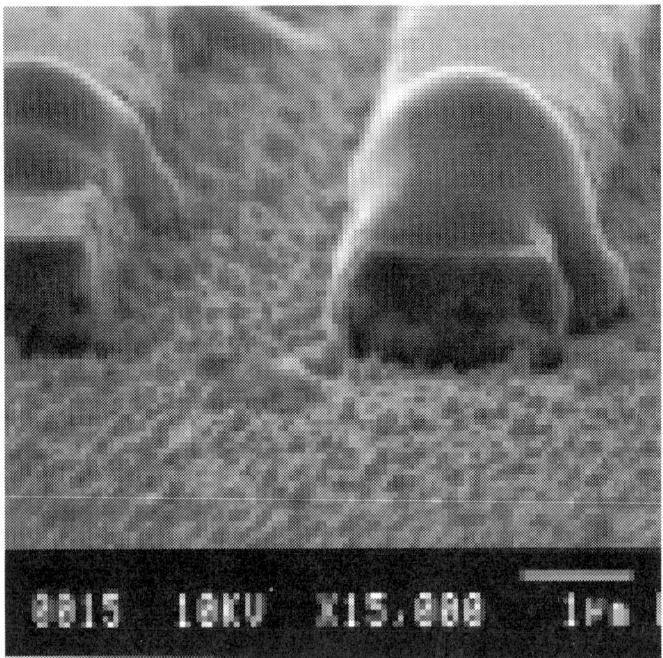

FIGURE 1.2. This figure shows a SEM image of a GaAs wafer. The surface consists of metal deposited via evaporation on a substrate after an RIE etch process. The surface appears rough due to a poor process. The problem is to come up with a quantitative measurement for the visual appearance of surface roughness. This is provided in chapter 5

process or a deposition process. Figure 1.2 shows an example of a texture that was produced by a poor deposition process. The problem here is to characterize the surface texture in a meaningful quantitative manner, which is a prevalent problem in many inspection tasks.

The reality of the situation is that most descriptions of defects are vague and subjective. For instance, inspectors use highly subjective methods to assess textural properties such as roughness.

Though the identification, description and classification of defects and anomalies is crucial to process control and automated inspection, there is unfortunately no *standardized* scheme to describe defects and anomalies. Faced with an absence of standardized symbols, inspection personnel in the semiconductor industry are forced to devise peculiar jargon to describe various textures that are created during the manufacturing process. Most of the features used are highly subjective, and terminology varies considerably.

Figure 1.3 illustrates some concrete jargon that is popularly used to describe different kinds of anomalies and defects arising during wafer pro-

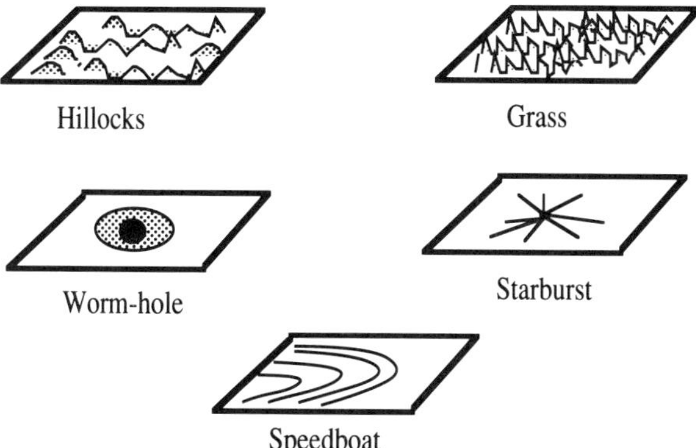

FIGURE 1.3. Illustrating the jargon used by the semiconductor industry to describe different kinds of visual defects arising during the manufacturing process

cessing. Such a scheme is ad-hoc, and a more scientific scheme is desirable. For instance, the Hewlett Packard Photolithography Advisor [35] defines an *orange peel defect* to occur when 'the photoresist has a wrinkling effect similar to the skin of an orange.' The system also asks the user to make a distinction between light and heavy orange peel – which is very subjective. Figure 1.4 illustrates a possible instance of an orange peel image. Another type of defect is illustrated in figure 1.5, which shows a 240x240 pixel image of a resist gel defect, obtained from [32]. The examples presented so far raise two important issues – what kinds of scientific descriptions are possible for these defects, and what computational methods can be used to characterize such defects quantitatively.

1.4.3 THE DESIRED SOLUTION

Given the above problems in automating the recognition and description of defects, there is need for a solution. What is really required is a symbolic representation scheme that is capable of providing a scientific description scheme for texture, and techniques for providing quantitative measurements related to texture.

This entails the design of a standardized description scheme for textures. The advantages of a standardized scheme are that it will allow for a description of new defects that may arise in the future and that cataloging the defects will become systematic [104].

Hence, *both* quantitative and qualitative techniques are required for the description and measurement of texture. This is precisely the focus of the book. Thus, the book fulfills a very practical need imposed on computer vision systems.

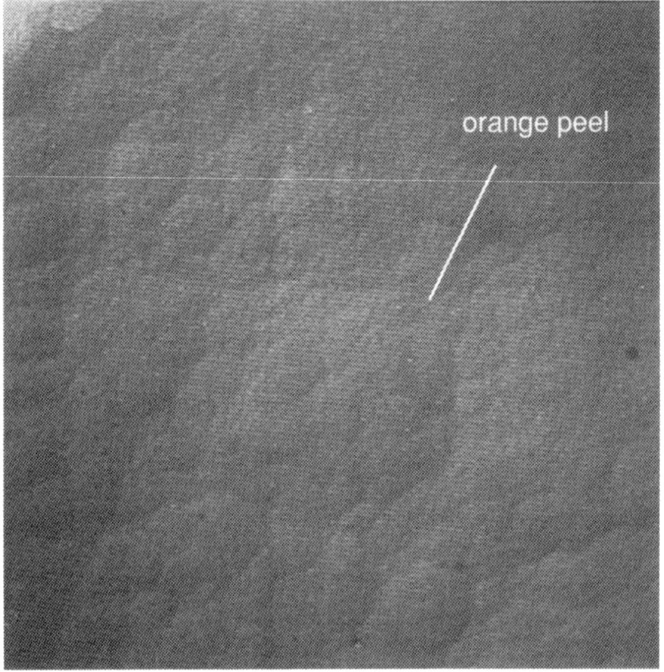

FIGURE 1.4. The original image of the orange peel defect. The problem here is to come up with a quantitative measure for the amount of wrinkle present on the wafer. In chapter 2 we give one plausible measure

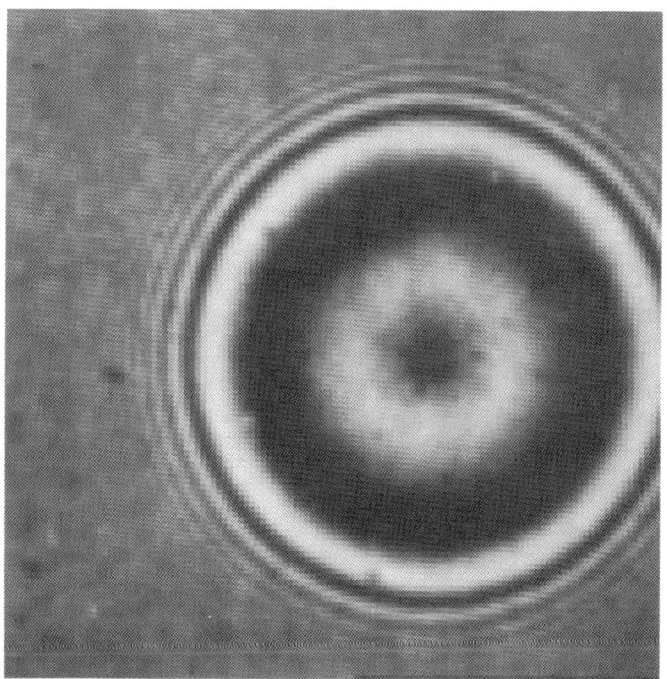

FIGURE 1.5. An image of resist gel defect, obtained from [32, pg. 254] (Reproduced with permission from *Integrated Circuit Mask Technology* by D. J. Elliott, pg. 254, ©1985 McGraw Hill). In chapter 3 we analyze this defect, and come up with a symbolic description.

1.5   A taxonomy for texture

Most of the research done in texture is concerned with either statistical or structural approaches. The goal of the *statistical approach* is to estimate parameters of some random process, such as fractal Brownian motion, or Markov random fields, which could have generated the original texture. Alternately, measures defined on the co-occurrence matrix have been used for statistical descriptions of texture. For the sake of nomenclature, let us refer to those textures for which a statistical model is appropriate as **disordered textures**.

Structural approaches try to describe a repetitive texture in terms of the primitive elements and placement rules that describe geometrical relationships between these elements. We will refer to those textures that are amenable to a structural description as being **strongly ordered textures**.

An important category of textures which can neither be modeled statistically nor structurally is that of **weakly ordered**, or oriented textures. Such textures are characterized by a dominant local orientation at each point of the texture, which can vary arbitrarily.

The above three kinds of textures are found widely in both natural and man-made objects, and will be the subject of study in this book. (See table 6.1 in chapter 6 for a classification of a variety of textures into these categories).

Examples of the different types of textures are shown in figure 1.6. **Strongly ordered textures** are those which exhibit either a specific placement of some primitive elements (like a brick wall, figure 1.6a) or are comprised of a distribution of a class of elements (such as grains within a matrix, figure 1.6b). **Weakly ordered textures** are those that exhibit some degree of orientation specificity at each point of the texture (figure 1.6c). **Disordered textures** (figure 1.6d) are those that show neither repetitiveness nor orientation and may be described on the basis of their roughness.

1.6   Outline

Texture has been an active area for research for over two decades [46, 125]. There is a considerable literature on visual texture, which indicates there are several kinds of texture that should be considered as separate visual phenomena. Hence each chapter in this book deals with a different aspect of texture.

The unifying theme of this book is to provide symbols that will serve as a descriptive vocabulary for textures. One important class of textures that has received attention only recently is flow-like texture [65, 106], which forms an important part of this book. Hence, the book begins with the

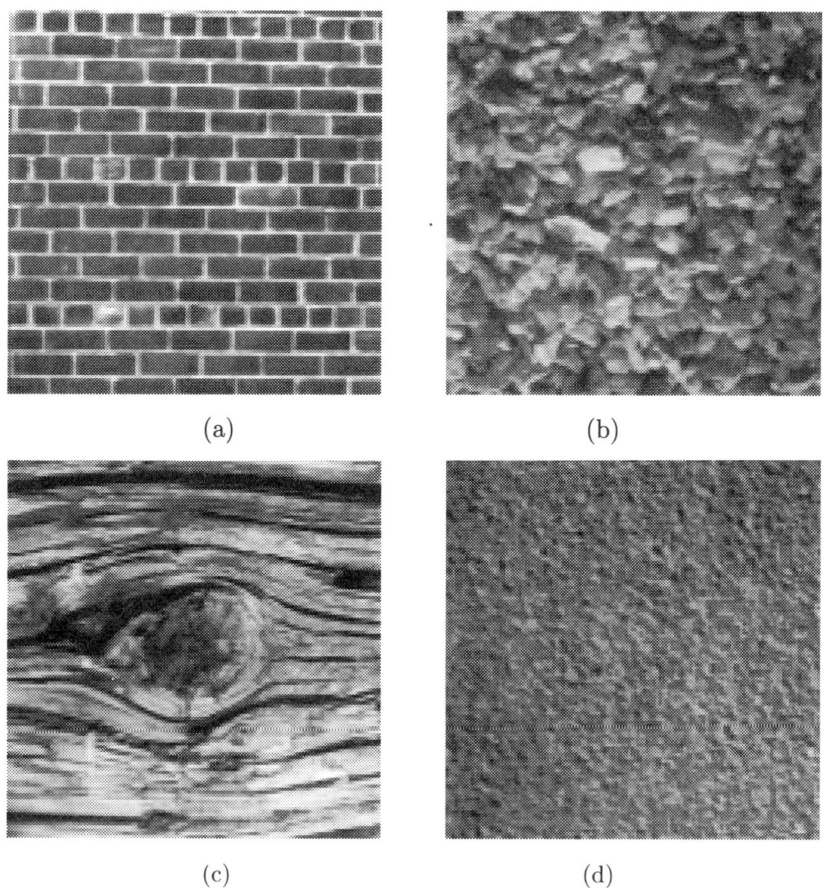

(a)　　　　　　　　　　(b)

(c)　　　　　　　　　　(d)

FIGURE 1.6. Illustrating the taxonomy for texture; **(a)** A brick wall, (d95 in Brodatz [19]). **(b)** Pebbles (d5) **(c)** Knot in a piece of wood (d72) **(d)** Surface of cork (d4)

analysis of oriented or flow-like textures.

The organization of the book is as follows.

CHAPTER 2

In chapter 2, we analyze the problem of computing oriented texture fields. We consider visual textures that are comprised of flow-like patterns such as wood grain. These textures are characterized by local selectivity of orientation, which can vary arbitrarily over the entire image. In other words, the texture is anisotropic. Every point in the image is associated with a dominant local orientation, and a local measure of the coherence or degree of anisotropy of the flow pattern. One way of visualizing oriented textures is to think about the image intensity surface as being comprised of ridges, whose direction *and* height can vary continuously. We define the *orientation field* of a texture image to be comprised of two images, called the *angle image* and *coherence image*. The angle image captures the dominant local orientation at each point in the texture in terms of an angle, and the coherence image represents the degree of anisotropy at each point in the texture. The term orientation *field* is used because it is closely related to a velocity flow field, where at each point in space, a fluid element can have some velocity, which is composed of a magnitude and direction. However, the analogy is not complete, as we shall show later. The major contributions of this chapter are to present optimum methods for the computation of the angle and coherence images, and to present the results of several experiments that illustrate the importance of these images in different contexts.

Kass and Witkin [65] have developed an algorithm based on Laplacian of Gaussian filters for estimating the orientation of texture patterns. We present another algorithm based on the gradient of the Gaussian. The most important aspect of the new algorithm is that it is optimal in estimating the local orientation of an oriented texture. The new algorithm incorporates a better scheme for computing the coherence of the flow field. Furthermore, some attractive features of the new algorithm are that it requires less computation and has better signal to noise characteristics since it incorporates fewer derivative operations.

We present optimum methods for the computation of the angle and coherence images, and illustrate the use of these images in different contexts [106]. The computation of the angle and coherence images can be regarded as the first stage in deriving a symbolic description of an oriented texture.

CHAPTER 3

In chapter 3, we extend the work in chapter 2, and complete the process of signal to symbol transformation for oriented textures. The objective of this chapter is to analyze a specific kind of texture, called oriented or flow-like texture, and to propose a solution for the signal to symbol transformation in this domain. Our approach to the problem of oriented texture description is to view it as a two stage process. The first stage is concerned with extracting an orientation field from the raw image, as described in chapter 2. The second stage is concerned with performing computations on these intrinsic images in order to derive a qualitative description of the orientation field, and hence the underlying texture. We present a method to analyze a given orientation field based on the geometric theory of differential equations [76, 7, 51, 6].

This theory involves the notion of a *phase portrait*, which represents the solution of a system of differential equations in a qualitative fashion. For a linear 2D system, only a finite number of qualitatively different phase portraits are possible. This allows a parametric representation of different types of phase portraits, and their classification is possible based on the notion of qualitative equivalence. The basic idea is to view a given texture flow pattern as being comprised of piecewise linear flows, and to describe each linear flow by means of an equivalent phase portrait.

We pose the problem as one of non-linear least squares optimization, which is solved using the Levenberg-Marquardt technique. We also review the equivalence between the differential analysis of fluid motion and the geometric theory of differential equations. The interesting part of this analysis is that these two representations complement each other.

The main result of this chapter is that the theory of differential equations can be successfully used to describe flow-like textures both qualitatively and quantitatively. An attractive feature of this method is that it is completely domain independent, and makes no assumptions about the kind of texture that may be present. We present the results of applying the method to several real texture images.

CHAPTER 4

In chapter 4, we examine *strongly ordered textures*, also known as repetitive textures. These textures can be modeled by a primitive element that is replicated according to certain placement rules. In keeping with the spirit of this book, this chapter provides a vocabulary for primitives as well as placement rules.

Very few researchers have attempted to give an explicit symbolic description for repetitive textures. We present a precise technique for providing symbolic descriptions, based on the equivalence classes of the frieze groups and the wallpaper groups.

The aim of this chapter is to take a critical look at the research done in the structural analysis of textures, with an eye for ultimately deriving accurate symbolic descriptors for ordered textures. Thus, we provide directions for future research in ordered textures based on existing research.

CHAPTER 5

In chapter 5 we analyze disordered textures. **Disordered textures** are those that show neither repetitiveness nor orientation. There are several methods for analyzing disordered textures, based on features such as roughness, edgeness, entropy, contrast, and fractal dimension. In this chapter we briefly review some of these techniques, and focus on methods to measure the fractal dimension of an image.

CHAPTER 6

The major portion of the book is concerned with analyzing models for specific kinds of texture, namely disordered, strongly ordered and weakly ordered. We now present a scheme which uses the descriptors for these types of texture as primitives, and constructs descriptions of more complex textures, called **compositional textures**. In order to perform this task, we need rules of composition, which will allow the construction of arbitrary textures. This chapter defines certain rules of composition which are useful in the analysis of textures.

The creation of the class of compositional textures allows, in principle, a complete classification of all the textures from an album such as Brodatz [19].

2

Computing oriented texture fields

2.1 Introduction

This chapter deals with visual textures that are comprised of flow-like patterns such as wood grain. These textures are characterized by local selectivity of orientation, which can vary arbitrarily over the entire image. In other words, the texture is anisotropic. Every point in the image is associated with a dominant local orientation, and a local measure of the coherence or degree of anisotropy of the flow pattern. One way of visualizing oriented textures is to think about the image intensity surface as being comprised of ridges, whose direction *and* height can vary continuously. We define the *orientation field* of a texture image to be comprised of two images, called the *angle image* and *coherence image*. The angle image captures the dominant local orientation at each point in the texture in terms of an angle, and the coherence image represents the degree of anisotropy at each point in the texture. The term orientation *field* is used because it is closely related to a velocity flow field, where at each point in space, a fluid element can have some velocity, which is composed of a magnitude and direction. However, the analogy is not complete, as we shall show later. The major contributions of this chapter are to present optimum methods for the computation of the angle and coherence images, and to present the results of several experiments that illustrate the importance of these images in different contexts.

Kass and Witkin [65] have developed an algorithm for estimating the orientation of texture patterns. Their method can be used to decompose the oriented pattern into a flow field, describing the direction of anisotropy, and describing the pattern independent of changing flow direction. The algorithm developed by Kass and Witkin [65] for estimating the texture orientation was based on Laplacian of Gaussian filters. This chapter presents another algorithm based on the gradient of the Gaussian. The most important aspect of the new algorithm is that it is optimal in estimating the local orientation of an oriented texture. This result is an improvement over Kass and Witkin's scheme [65] as it makes no assumptions about the nature of the oriented texture (such as producing a zero mean filter output after convolving with the Laplacian of Gaussian). The new algorithm incorporates a better scheme for computing the coherence of the flow field. Furthermore, some attractive features of the new algorithm are that it requires less computation and has better signal to noise characteristics since it incorporates fewer derivative operations.

2.2 Background

The objective of this chapter is to obtain an orientation field from a given texture. Based on the philosophy of Barrow and Tenenbaum [8], the orientation field may be viewed as an intrinsic image, and represents the first computational stage in analyzing oriented textures. The orientation field is computed using domain-independent processing, and constitutes a more abstract description of the image than raw intensity values. In the final section, we show how predicates defined on the intrinsic image give rise to meaningful results in appropriate domains.

Many of the classic approaches to texture identification and description are based on co-occurrence matrices [46]. The co-occurrence matrix is defined as follows. Consider the second-order joint conditional probability density function, $f(i, j \mid d, \theta)$. For a given θ and d, $f(i, j \mid d, \theta)$ represents the probability of going from gray level i to gray level j, given that the intersample spacing is d and the direction of the intersample spacing is θ. For a given value of d and θ we can generate a matrix which represents the estimated second order joint conditional probability density function, which is known as the *gray level co-ocurrence matrix* [46]. There are many problems associated with the use of the co-occurrence matrices. Firstly, a co-occurrence matrix has to be computed for different values of d and θ, and then the appropriate texture measure has to be computed from each co-occurrence matrix. Secondly, co-occurrence matrices are inappropriate to describe oriented textures since directionality can vary continuosly in an oriented texture. This is because the co-occurrence matrix is defined only for *globally* discrete values of θ, and is hence incapable of estimating local variations in orientation. In this chapter we present a more powerful tool to analyze oriented textures.

Aloimonos and Swain [2, 4] presented an algorithm for determining the orientation of surfaces from the distortion of textures under perspective projection. The algorithm assumes that the scene surface is covered with uniformly repeated texture elements. It would be interested to explore how the algorithm would work when presented with perspective distortion of texture flow patterns. Coleman and Jain [22] extend the work on shape from shading to include variations in the albedo. Essentially, the variations in albedo form a texture pattern and their shape from shading algorithm compensates for the effect of texture on the shape from shading algorithm.

Julesz and Bergen [61] introduced the notion of textons, which are features that are extracted by the preattentive visual system. Textons are visual primitives such as blobs and terminations from the primal sketch theory of Marr [83] and crossings of line segments. Textons have specific properties such as color or orientation. This chapter addresses the extraction of the orientation of primitive features.

Kanatani [62] used integral geometry to develop an algorithm that determines surface orientation from the intersections the image scan lines with curves of texture. In principle, the texture orientation field could be used as a source for the curves of texture used in the work of Kanatani.

Tomita, Shirai, and Tsuji [115] presented an algorithm for extracting texture elements and the placement rules from images. The texture orientation field algorithm presented in this chapter could be used as a preprocessing stage to extract the texture pattern. The notion of texture coherence and angle could guide the formulation of the placement rules that define the repetition pattern for the texture elements.

Vilnrotter, Nevatia, and Price [120] used a similar approach based on computing the repetition of edge elements in a texture. Their work is directed towards computing an interpretation of the texture. They compute edge repetition arrays, which are similar to co-occurrence matrices, from the edge detector outputs. The algorithm presented in this paper does not require the explicit computation of the edge repetition structure as the coherence and angle intrinsic images play a similar role, but the algorithm presented in this chapter is restricted to flow-like textures that can be modeled by coherence and angle alone.

Witkin [127] presents an algorithm for computing surface orientation from texture the possible directions for the tangents of markings on the surface are equally likely, and surface orientation and tangent direction are independent. These are the assumptions of isotropy and independence, without assuming that the texture pattern is regular. The only assumptions are that all possible surface orientations are equally likely. If there is a flow-like texture, then all possible tangent directions are not equally likely. It is still important to run experiments to determine the effect of violating isotropy.

Zucker and Cavanaugh [135] have performed experiments with subjective figures in texture discrimination. The test image is called the Ehrenstein illusion. The two versions of the illusion, called the pattern and the control, should have different texture orientation field structures which could contribute to part of an explanation for the subjective effects. We present results related to the illusion in this chapter.

Strassman has analyzed the style or texture of brush strokes in Japanese paintings [112]. He tries to create a realistic computer paint brush by modeling the compound interactions between brush, paint and paper. It would be interesting to determine whether the algorithms presented in this chapter could be used to automatically determine the parameters of the brush model specified by Strassman.

Bovik et al [17] have developed a multichannel approach to texture analysis using Gabor filters [29]. Textures are modeled as signals of band-limited spatial frequency. Variations in the amplitude and phase repsonse of the

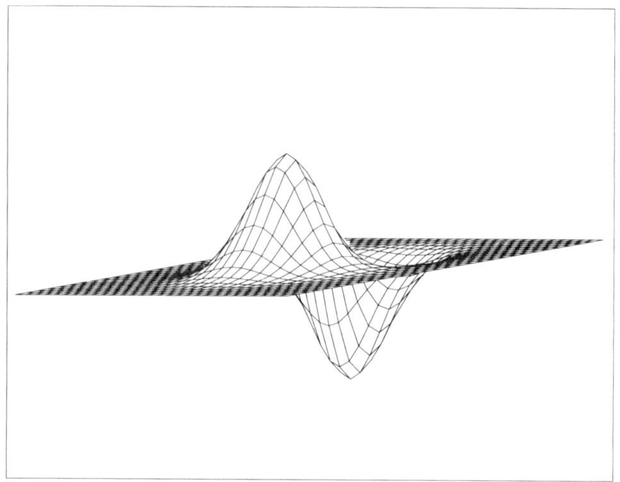

FIGURE 2.1. The Fourier transform of the first derivative of a Gaussian filter is tuned to detect oriented textures at a particular orientation and wavelength.

filter are analyzed to segment texture mosaics.

2.3 Oriented Texture Fields

The algorithm for estimating the orientation of a texture field presented in this chapter is based on the gradient of a Gaussian filter. The approach is similar to the work of Kass and Witkin [65], but involves fewer derivative operations and hence has a better signal to noise ratio. Figure 2.1 is a plot of the magnitude of the Fourier transform of the first derivative of a Gaussian filter. In two dimensions, the Fourier transform of the gradient of a Gaussian filter has two lobes oriented along a line passing through the origin in frequency space. The frequency spectrum of a sine-wave grating has two δ-functions at equal distances from the origin along a line in frequency space. An oriented pattern will have a dominant frequency component and the response of a gradient of Gaussian filter can be tuned to this dominant component.

The Gaussian smoothing filter has the impulse response given by

$$g(x) = e^{-x^2/2\sigma^2} \tag{2.1}$$

and Fourier transform

$$G(\omega) = \sqrt{2\pi}e^{-\omega^2/2\nu^2}, \qquad \nu = \frac{1}{\sigma}. \tag{2.2}$$

The magnitude of the Fourier transform of the first derivative of a Gaussian filter is

$$|i\omega G(\omega)| = |\omega|\sqrt{2\pi}\sigma e^{-\omega^2/2\nu^2}. \tag{2.3}$$

The location of the maximum response of the first derivative of a Gaussian filter is easily determined. Let the Fourier transform of the first derivative of a Gaussian be

$$H(\omega) = \frac{\sqrt{2\pi}}{\nu}\omega e^{-\omega^2/2\nu^2}. \tag{2.4}$$

The location of the maximum can be determined by setting the derivative of $H(\omega)$ to zero. From this analysis,

$$H'(\omega) = \frac{\sqrt{2\pi}}{\nu}\left[e^{-\omega^2/2\nu^2} - \omega\frac{2\omega}{2\nu^2}e^{-\omega^2/2\nu^2}\right] \tag{2.5}$$

$$= \frac{\sqrt{2\pi}}{\nu}\left[1 - \frac{\omega^2}{\nu^2}\right]e^{-\omega^2/2\nu^2}. \tag{2.6}$$

It is easy to see that the maximum response occurs at a distance of $\omega = \nu = 1/\sigma$ from the origin the frequency space. In two dimensions, the maximum response is oriented along a line passing through the origin in frequency space. If the orientation of the line is θ, then the filter is the derivative of the Gaussian in the direction of θ; in other words, the normal derivative:

$$\frac{dg}{d\hat{\mathbf{n}}} = \nabla g \cdot (\cos\theta, \sin\theta). \tag{2.7}$$

The orientation of the filter is adjusted to achieve the maximum response for the underlying visual texture. In implementation, the orientation of the maximum response is determined by first computing the gradient after smoothing with a Gaussian filter, and then processing the gradient vector field as follows. The orientation of the gradient is the direction of maximum response to within a sign reversal.

There are five steps to estimating the local orientation of the texture field:

1. Smooth the image with a Gaussian filter;

2. Compute the gradient of the smoothed image;

3. Find the local orientation angle using the inverse tangent;

4. Average the local orientation estimates over a neighborhood;

5. Compute a measure of the coherence (the degree of flow-like texture) of the pattern.

The first two steps are standard algorithms in edge detection [21, 109]. These steps are discussed in more detail in sections 2.5 and 2.7. The gradient of Gaussian operator is near-optimal for edge detection, as shown by Canny [21]. This notion of optimality for edge detection can be carried over to the analysis of oriented textures, since the two problems have similar initial processing stages. Kass and Witkin use a laplacian of Gaussian operator as the oriented filter. Clearly, the gradient of Gaussian involves one less differentiation, and will have better performance in the presence of noise.

Since the smoothing of the image has to be performed at a given scale, this aspect will be explored in this chapter in greater detail. As we shall show, there is no one scale that can be used to describe all textures, and in fact, the choice of scale can dramatically affect the description of a texture. Thus, the choice of scale is dependent to a large extent on what feature sizes one would like to focus on, and this is dictated by prior goals of processing.

The third step can be easily performed using the inverse tangent function. In most edge detection algorithms, the angle of the edge covers the full range of the unit circle; but in estimating the orientation field, it is necessary to reflect orientation vectors that lie along the same line to a canonical orientation. In edge detection, the gradient angle is computed using the arctangent function of two arguments [109]; but if this function were used to compute the orientation angle in this problem, then there would not be a unique angle for each texture orientation. The texture field orientations would have to be post-processed to reflect orientation vectors into a canonical range as is apparently done by Kass and Witkin [65]. The texture orientation angle can be computed using the inverse tangent function of one argument which results in angles in the range $(-\pi/2, \pi/2)$ and the representation of texture orientation is unique. This technique is simpler than the rescaling operation that must be performed as part of the orientation estimation algorithm of Kass and Witkin [65], but does not work as well as expected for reasons explained in section 2.4.4.

After computing the local orientation of the texture field, the orientation estimate must be smoothed to compute the average orientation over a neighborhood of significant size. In order to perform this, we present a best estimate for the dominant local orientation in section 2.4.1. Let σ_1 be the width of the first Gaussian smoothing filter used to compute the local texture orientation and σ_2 be the width of the second smoothing filter used to average the orientation estimate. The averaging filter must be large enough to average the orientation from several of the local estimates and so $\sigma_2 \gg \sigma_1$, but the averaging filter should be smaller than the distance over which the orientation of the texture field undergoes major changes; in other words, the averaging filter should not blur changes in the texture field.

The *coherence* is the oriented texture pattern is the degree to which the local orientation estimates computed before averaging agree. The essential idea in measuring coherence is to project the orientation vectors in some neighborhood onto a representative orientation vector from the neighborhood and normalize the result. If the orientations are coherent, then the normalized projections will be close to one; but if the orientations are not coherent, then the projections will tend to cancel and produce a result close to zero.

2.4 Experimental Methods

Test images were digitized using a CCD camera from photographs of textures available in published sources [118, 19]. The image is first smoothed with a Gaussian filter. The sizes of the filter are indicated in the figures. The algorithm for computing the filter coefficients can be found in [109]. To make this operation efficient, we have implemented the convolution in a separable manner [108]. The gradient of the smoothed image is computed using finite differences. Let $G_x(i, j)$ and $G_y(i, j)$ be the x and y components of the gradient vector at point (i, j). In order to calculate the orientation at a point, one needs to combine the gradient orientations in the neighborhood of that point. As Kass and Witkin pointed out [65], one cannot smooth the gradient vectors, as they tend to cancel each other out at intensity ridges. There are several methods that one can use to avoid such cancellation, and we shall present a best estimate for the dominant orientation within a neighborhood.

2.4.1 A BEST ESTIMATE FOR DOMINANT LOCAL ORIENTATION

Consider a set of line segments, as shown in figure 2.2. The problem that needs to be solved is – how does one determine the dominant orientation of this set of segments?

One may think that it is sufficient to sum up these line segments vectorially and find the resultant direction. However, this will not work for two reasons. Firstly, any given line segment does not have a unique direction, since it could be taken to point either in the direction θ or $\theta + \pi$. Secondly, even if the line segments were assigned directions, there is the danger that segments pointing in opposite directions will cancel each other out, instead of influencing the choice of dominant orientation as they should. One way to tackle this problem is as follows.

Assume that the segments are indexed by the subscript i, where i ranges from 1 to N, the number of segments. Consider a line oriented at an angle θ as shown. Let the jth segment subtend an angle θ_j. The next few steps will show that it does not matter what sense this angle is taken in, ie it is immaterial as to what vector direction one chooses for the line segment.

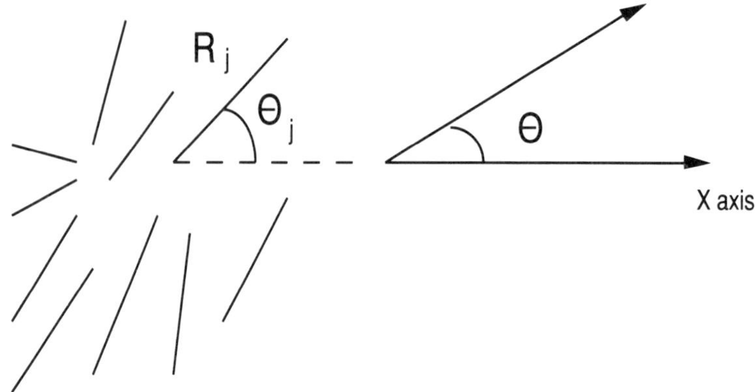

FIGURE 2.2. Illustrating the method to compute the dominant orientation in a group of line segments

Let R_j be the length of the jth segment. The projection of this segment onto the line is $R_j cos(\theta_j - \theta)$. Consider the sum of the absolute value of all such projections,

$$S_1 = \sum_{j=1}^{j=N} \| R_j cos(\theta_j - \theta) \| \qquad (2.8)$$

S_1 varies as the orientation of the line θ is varied. That value of θ which maximizes S_1 is the dominant orientation of the given set of line segments. Thus, one can evaluate the dominant orientation by maximizing S_1 with respect to θ. Since the absolute value function is not differentiable everywhere, one can equivalently maximize the sum S_2, where

$$S_2 = \sum_{j=1}^{j=N} R_j^2 cos^2(\theta_j - \theta) \qquad (2.9)$$

Here we have taken the sum of the square of the projections. Differentiating equation 2.9 with respect to θ we get

$$\frac{dS_2}{d\theta} = - \sum_{j=1}^{j=N} 2R_j^2 cos(\theta_j - \theta)sin(\theta_j - \theta) \qquad (2.10)$$

Equating $dS_2/d\theta$ to zero in order to obtain an extremum, we get from the above equation,

$$\sum_{j=1}^{j=N} R_j^2 sin2(\theta_j - \theta) = 0 \qquad (2.11)$$

$$\sum_{j=1}^{j=N} R_j^2 sin2\theta_j cos2\theta = \sum_{j=1}^{j=N} R_j^2 cos2\theta_j sin2\theta \qquad (2.12)$$

Hence

$$tan2\theta = \frac{\sum\limits_{j=1}^{j=N} R_j^2 sin2\theta_j}{\sum\limits_{j=1}^{j=N} R_j^2 cos2\theta_j} \tag{2.13}$$

Let $\hat{\theta}$ be the value of θ which satisfies equation 2.13. Instead of finding the sum S_2 at different orientations, equation 2.13 tells us in a single computation the orientation $\hat{\theta}$ that maximizes S_2 and is hence the dominant direction of the pattern of line segments. Since $\hat{\theta}$ maximizes S_2, we call it the *best estimate for dominant local orientation*. That $\hat{\theta}$ indeed maximizes S_2 will be proved shortly.

Equation 2.13 has an interesting interpretation. Consider the line segments to lie in the complex plane, each segment being represented by $R_j e^{i\theta_j}$, where R_j is the length of the segment and θ_j is its direction. Now square all the segments, which have been represented as complex numbers. Thus each segment will give rise to a term of the form $R_j^2 e^{2i\theta_j}$. If one sums these numbers, the resulting complex number has an orientation α, with respect to the x-axis, given by

$$tan\alpha = \frac{\sum\limits_{j=1}^{j=N} R_j^2 sin2\theta_j}{\sum\limits_{j=1}^{j=N} R_j^2 cos2\theta_j} \tag{2.14}$$

This equation is the same as equation 2.13.

Using the above interpretation, one can show that $\hat{\theta}$ derived from equation 2.13 indeed maximizes S_2. In order to show this, one must show that $d^2 S_2/d\theta^2$ is negative at $\theta = \hat{\theta}$. From equation 2.10 we get

$$d^2 S_2/d\theta^2 = -2 \sum\limits_{j=1}^{j=N} R_j^2 cos(2\theta_j - 2\theta) \tag{2.15}$$

If we show that the summed quantity is positive we are done. In order to do this, we must assume that the texture has only one dominant local orientation. In this case, the term $R_j^2 cos(2\theta_j - 2\theta)$ represents the projection of the squared vector onto the line oriented at 2θ. The sum of such projections is positive when the texture has only one dominant local orientation. Thus, we have proved that equation 2.13 provides us with a best estimate of the local orientation within a neighborhood, in the sense that it maximizes the norm in equation 2.9.

Another interesting point of comparison can be made as follows. Equation 2.13 can be regarded as a smoothing operation on the image, where a box filter (all coefficients identical) is used for smoothing. Kass and Witkin use a Gaussian envelope to perform the smoothing, whereas we have used a box filter for smoothing. We found that the box filter gave better looking results, as shown in figure 2.9.

In order to use equation 2.13 to estimate the orientation at each point in the image, one can consider the gradient vector having components G_x and G_y to represent the line segments of figure 2.2. Consider the vector in the complex plane formed by combining G_x and G_y as $(G_x + iG_y)$. Let the gradient vector at point (m, n) in the image have the polar representation $R_{mn}e^{i\theta_{mn}}$. Thus, the estimate of the dominant orientation θ in an NxN neighborhood of the image would be given by

$$\theta = \frac{1}{2}tan^{-1}\left(\frac{\sum_{m=1}^{m=N}\sum_{n=1}^{n=N} R_{mn}^2 sin2\theta_{mn}}{\sum_{m=1}^{m=N}\sum_{n=1}^{n=N} R_{mn}^2 cos2\theta_{mn}}\right) \tag{2.16}$$

The estimated orientation angle at (m, n) is then $\theta_{mn} + \pi/2$, since the gradient vector is perpendicular to the direction of anisotropy. When we refer to the *scale* of the smoothing filter to combine information about the gradient vectors, we mean the size N of the neighborhood over which the estimate is obtained, as in equation 2.16. In all our results we display the estimated orientation angle overlayed on the original image.

2.4.2   DERIVATION USING THE MOMENT METHOD

The derivation presented above is for the two dimensional case, and is equivalent to the *moment method*, which holds for N dimensions.

Problem: Given vectors $\mathbf{v_1}, ... \mathbf{v_K}$, estimate the average orientation when the sign of $\mathbf{v_i}$ is ignored.

Solution: The average orientation is given as the principal axis of the moment tensor

$$\mathbf{M} = \sum_{i=i}^{i=K} \mathbf{v_i}\mathbf{v_i}^T \tag{2.17}$$

corresponding to the maximum principal value, where $(.)^T$ designates the transpose.

Proof: Let $\bar{\mathbf{v}}$ be the unit vector along the average orientation to be estimated. The projection of vector $\mathbf{v_i}$ along this orientation has a magnitude $|(\mathbf{v_i}, \bar{\mathbf{v}})|$, where $(.,.)$ denotes the inner product. Hence the average orientation $\bar{\mathbf{v}}$ can be estimated by maximizing

$$J = \sum_{i=1}^{i=K}(\mathbf{v_i}, \bar{\mathbf{v}})^2 \tag{2.18}$$

In terms of the moment tensor $\mathbf{M}$, this expression becomes

$$J = (\bar{\mathbf{v}}, \mathbf{M}\bar{\mathbf{v}}) \tag{2.19}$$

This quadratic form in $\bar{\mathbf{v}}$ is maximized under the constraint $\mid \bar{\mathbf{v}} \mid = 1$. The solution is given by the eigenvector $\bar{\mathbf{v}}$ of $\mathbf{M}$ corresponding to the maximium eigenvalue, a well known result in linear algebra.

From this result, the following observations can be made. Both $\bar{\mathbf{v}}$ and $-\bar{\mathbf{v}}$ give rise to the same maximum value for J. Similarly, replacing any $\mathbf{v_i}$ by $-\mathbf{v_i}$ will not alter J, as the latter is quadratic in $\mathbf{v_i}$. Finally, the above proof is independent of the dimensionality of the vector space corresponding to $\mathbf{v_i}$.

2.4.3 SQUARING THE GRADIENT VECTORS: KASS AND WITKIN'S SCHEME

We now discuss an alternate method, used by Kass and Witkin [65] in order to smooth the gradient vector field. Consider the vector in the complex plane formed by combining G_x and G_y as $(G_x + iG_y)$. Let this vector have the polar representation $Re^{i\theta}$. The square of this vector is $R^2e^{2i\theta}$. Consider the vector $Re^{i(\theta+\pi)}$, which points opposite to $Re^{i\theta}$. The square of this vector is $R^2e^{2i\theta+2\pi} = R^2e^{2i\theta}$. Hence, squaring gradient vectors that point in opposite directions makes them reinforce each other. This is the basis of the first scheme for combining gradient orientations, and has been used by Kass and Witkin [65].

Let $J(i,j)$ denote the squared gradient vector at (i,j). The x component of J is $J_x(i,j) = G_x(i,j)^2 - G_y(i,j)^2$ and the y component is $J_y(i,j) = 2G_x(i,j)G_y(i,j)$. J_x and J_y are computed from the gradient of the image in this manner. The next step is to smooth J_x and J_y in order to average the orientation estimates over a neighborhood. This is again done using Gaussian filters, as in the first stage. Let $J_x^\star(i,j)$ and $J_y^\star(i,j)$ represent the smoothed squared gradient vector at (i,j). Let θ_{ij} be defined by the equation

$$\theta_{ij} = \tan^{-1}\left(\frac{J_y^\star(i,j)}{J_x^\star(i,j)}\right)/2, \tag{2.20}$$

where the arctangent is computed using two arguments, and lies in the range $[0, 2\pi)$. The division by 2 occurs because the original gradient vector was squared. The estimated orientation angle at (i,j) is then $\theta_{ij} + \pi/2$, since the gradient vector is perpendicular to the direction of anisotropy.

Kass and Witkin arrived at the above result by making several approximating assumptions which veil its importance. Firstly, they use the mean of the variance of the output of the directional filter as an estimate of dominant orientation. Their analysis has been done in a very intuitive manner. However, the derivation presented earlier in this chapter employs the maximization of a norm, which is a more accurate representation of the problem at hand. Secondly, Kass and Witkin assume that the output of the directional filter is zero mean, which is unnecessary, because the derivation in section 2.4.1 does not require such assumptions. Finally, Kass and Witkin use a Gaussian envelope to perform a smoothing of the squared gradient vectors with the sole justification that 'Gaussian convolutions can be computed efficiently' [65, p. 366]. Again, this is an *ad hoc* scheme, based on intuition without rigorous justification. In fact, if efficient computation is the sole justification, then a box filter provides the most efficient filter for smoothing.

On the other hand, the method presented in section 2.4.1 is completely general, and the only assumption made is that there is a single dominant direction at each point in the image.

2.4.4 INVERSE ARCTANGENT

Instead of squaring the gradient vectors, a much simpler scheme is to use the arctangent of *one* argument, which returns $\tan^{-1} x$ in the range $(-\pi/2, \pi/2)$. This effectively takes vectors that point in opposite directions, and maps them onto the same direction, as is shown in figure 2.11. This is another way of ensuring that gradient vectors pointing in opposite directions actually reinforce each other instead of cancelling when smoothing is performed.

Using this second scheme, we compute θ_{ij}, defined by the following equation

$$\theta_{ij} = \tan^{-1}\left(\frac{G_y(i,j)}{G_x(i,j)}\right) \tag{2.21}$$

where the arctangent is computed using one argument, and lies in the range $(-\pi/2, \pi/2)$. We now smooth the array of θ_{ij} values using a Gaussian filter, as was done in the first stage. However, for this smoothing, we used a larger filter size than what was used for smoothing the image. Let $\theta_{ij}^{\star}$ denote the array of smoothed angle values. The estimated orientation angle at (i,j) is then $\theta_{ij}^{\star} + \pi/2$.

2.4.5 FLOW ORIENTATION COHERENCE

Let $\theta(x,y)$ denote the estimated orientation angle at point (x,y), found in the earlier step. Let $G(x,y)$ denote the gradient magnitude at point (x,y) in the image, as shown in figure 2.3. To find the coherence at point (x_0, y_0), consider the point (x_i, y_j), where i and j are chosen so that they fall within a window W of prescribed size around the point (x_0, y_0). Project the gradient magnitude $G(x_i, y_j)$ taken in the direction $\theta(x_i, y_j)$ onto the unit vector in the direction $\theta(x_0, y_0)$. This will be $G(x_i, y_j) \cos(\theta(x_0, y_0) - \theta(x_i, y_j))$. Compute the sum of the absolute value of such projections over all (i,j) values within the window. The absolute value is used in order to avoid the cancellation of vectors that point in opposite directions. Consider the

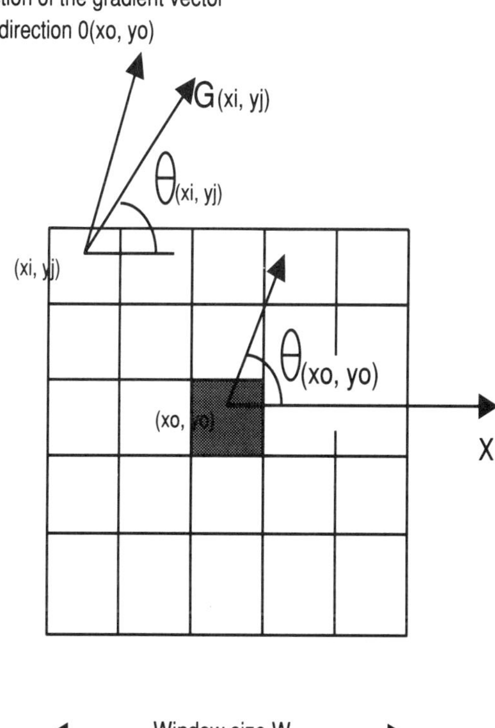

FIGURE 2.3. Illustration of the method used to compute the coherence of the texture flow field.

FIGURE 2.4.
The figure on the right shows the result of applying the coherence measure in equation 2.22 to the flow image on the left. *(Photograph courtesy M. Van Dyke)*

measure then defined by

$$\kappa = \frac{\sum_{(i,j)\in W} \| G(x_i, y_j) \cos(\theta(x_0, y_0) - \theta(x_i, y_j)) \|}{\sum_{(i,j)\in W} G(x_i, y_j)} \qquad (2.22)$$

This measure is related to the *dispersion* of directional data [82]. Assume that we have n data points, where the ith point P_i is at an angle θ_i, and lies on the unit circle. Let α be a fixed direction. Then the dispersion D for the n data points about the orientation α is defined by

$$D = \frac{1}{n} \sum_{i=1}^{n} (1 - cos(\theta_i - \alpha)) \qquad (2.23)$$

Thus if the measure D is generalized to data that does not necessarily lie on the unit circle, then it follows that $\kappa = 1 - D$. This shows that both the measure for orientation and coherence come from the same theory of statistics of directional data [82].

The result of applying this measure of coherence to the flow image is shown in figure 2.4. One can observe that the coherence values are low within the center of the image, and also around the four vortices in the corners. However, at other points in the image the coherence does not exhibit much variation.

One can obtain a better measure of coherence by weighting the quantity in equation 2.22 by the gradient magnitude at that point. Thus, the new measure of coherence is defined by

$$\rho = G(x_0, y_0) \frac{\displaystyle\sum_{(i,j)\in W} \| G(x_i, y_j) \cos(\theta(x_0, y_0) - \theta(x_i, y_j)) \|}{\displaystyle\sum_{(i,j)\in W} G(x_i, y_j)} \qquad (2.24)$$

The reason for weighting by the gradient magnitude is that we want the coherence to be high at points in the image which have high visual contrast, ie high gradients. The result of applying this measure of coherence to the flow image is shown in figure 2.12 (a). The coherence measure that we propose incorporates the gradient magnitude and hence places more weight on regions that have higher visual contrast.

Kass and Witkin propose the following coherence measure

$$\rho = (J_x^\star(i,j)^2 + J_y^\star(i,j)^2)^{1/2}/G^\star(i,j) \qquad (2.25)$$

where $J_x^\star(i,j)$ and $J_y^\star(i,j)$ have been defined in section 2.4.3 and $G^\star(i,j)$ is the smoothed gradient magnitude, obtained by smoothing the gradient magnitude with a Gaussian filter. We have used both coherence measures, and found that the coherence we propose in equation 2.24 gives better results, as is shown in figure 2.12.

2.4.6   THE EFFECT OF VARYING σ_1 ON THE ESTIMATE OF DOMINANT ORIENTATION

The scale corresponding to equation 2.1, σ_1, is the size of the Gaussian filter used to derive the gradient vectors as presented in section 2.4.1. We now provide a brief discussion of how varying σ_1 affects the estimate of dominant orientation in equation 2.16.

We shall restrict our analysis to the ideal case first. Consider again the example of a sine wave grating, say $S(x,y)$ which is an ideal oriented texture. According to the method presented in section 2.4.1, a best estimate of the dominant orientation at a point is provided by equation 2.16. This equation has embedded in it two smoothing operations: (1) smoothing with a Gaussian with variance σ_1; (2) a smoothing with a box filter of width σ_2. Let us consider the effect of varying σ_1 on equation 2.16, when σ_2 is constant.

Let $g(x,y)$ be the gradient vector at a point (x,y) of the function $S(x,y)$ $\otimes$ $G(\sigma_1, x, y)$, where G is the Gaussian filter. If we use a Gaussian filter of different size $G(\sigma_1', x, y)$, then it can be shown that the gradient vector at point (x,y) becomes $\frac{\sigma_1'}{\sigma_1} g(x,y)$, ie, there is a linear scaling. Furthermore, *all* gradient vectors get scaled by the same amount. The effect of scaling all gradient vectors by the same constant leaves the best estimate in equation 2.16 unchanged. Thus, for an ideal texture pattern the choice of

σ_1 to extract gradient vectors is immaterial. No matter what σ_1 is used, equation 2.16 provides the same estimate. In fact, this result holds for any texture that can be expressed as a linear combination of sine wave patterns.

In the general case, we are faced with non-linear textures (i.e. textures that cannot be expressed as a linear combination of sine wave patterns), as illustrated in the next section. For such cases, the above analysis no longer holds, and the choice of σ_1 will affect equation 2.16. However, the behaviour of the orientation estimation algorithm is not critically dependent on σ_1 or σ_2. It is only when these parameters are varied *significantly* that different responses could result, as shown in the next section. If one knows *a priori* the sizes of features in the texture pattern, then this knowledge can be used to estimate both the values of σ_1 and σ_2 needed. Normally one would use widely spaced σ's, as indicated in the edge detection theories of Marr and Hildreth [85] or Canny [21].

2.5 Experimental Results

Figure 2.5(a) shows a 240x240 image of secondary streaming induced by an oscillating cylinder [118, p. 23]. The cylinder, which is in the center of the image, is oscillated by a loudspeaker in a fluid. This pattern results when suspended glass beads are illuminated by a stroboscope, and consists of four vortices around which there is circulation. This texture is a useful test case because it exhibits orientation specificity at all possible angles, and is also symmetric. The results of applying the orientation estimation algorithm described in sections 2.4.1 and 2.4.5 are displayed in figures 2.5a and 2.5b. The orientation field is calculated for each point in the image, but is displayed in a sampled form in order to avoid clutter. No arrowheads are drawn for the line segments because there is an inherent ambiguity in the direction of the line segment: it could point in either one direction or exactly the opposite.

Two methods of presenting the orientation field are used in order to provide easy means for interpretation. In the first method, the orientation field is overlayed on the original image to aid comparison. The orientation at each point is represented by means of a line segment, where the direction of the segment corresponds to the dominant local orientation, and the length of the line segment is proportional to the strength of orientation (or coherence). In the second method, the orientation at each point in the image is encoded as the angle of orientation of a unit vector and is superimposed on the coherence or strength of orientation which is encoded as a gray value.

Figure 2.5(a) shows the result of applying the orientation estimation algorithm at a scale of $\sigma_1 = 5$ to compute the gradient vectors, and a scale of $\sigma_2 = 7$ to smooth the gradient vectors. The oriented segments are displayed at regularly sampled points overlayed on the original image. The segments capture the flow of the texture very well at each point, and orient themselves along the direction of flow. Hence the resulting pattern looks much like the original texture. Figure 2.5b shows the coherence of orientation (calculated using equation 2.24) encoded as an image. The coherence at each point is quantized into one of 256 gray levels, and then displayed as an image. The angle vectors are then overlayed on the coherence image. The coherence image shows how orientation specificity varies over the original image – the brighter points indicating strong coherence of flow. Note that the coherence at the center of the image is low, indicating that there is no dominant flow direction in that area. Thus, the coherence image combined with the overlayed angle image provides a good description of the underlying flow-like texture.

The filter sizes of 5 and 7 have been arbitrarily chosen at this point. In fact, figure 2.5(c) and 2.5(d) show the application of the orientation estimation algorithm at different scales to the same texture. The nature of the orientation field does not change significantly even though the filter sizes have been increased significantly. The results indicate that for the texture in question, the choice of scale does not play an important role in the description of the texture. However, this does not hold true for all textures, and the description of some textures change dramatically with changing scales, as illustrated in section 2.7.

Figures 2.6 and 2.7 show more results of applying our algorithm to flow textures scanned from an album of fluid motion [118]. The algorithm captures well the direction of anisotropy and the coherence at each point of the texture.

Figure 2.8a shows a knot in a piece of wood, scanned from a book of textures [19]. The orientation field swirls around the knot, though the fluid flow analogy is not entirely correct as there is no directionality to the vectors: it is an orientation field, not a vector field. The results of applying the orientation estimation algorithm described in sections 2.4.1 and 2.4.5 to the wood image are displayed in figure 2.8. The orientation vectors are drawn without arrow heads since the flow structure is non-directional.

2.5.1 COMPARING CALCULATIONS FOR ORIENTATION

Figure 2.9 shows a herringbone weave texture, obtained from[19]. Figure 2.9 compares the two schemes for measuring local orientation, as described by equations 2.16 and 2.20. Filter sizes used were $\sigma_1 = 5$ pixels and $\sigma_2 = 7$ pixels for both schemes. From a subjective estimate, equation 2.16 gives better results.

The use of equation 2.21 gives the result shown in figure 2.10a. This algorithm fails to respond well to flow patterns that are oriented horizontally. The problem with this approach is that gradient vectors having angles in the range $[\pi/2 - \epsilon, \pi/2)$ and $(-\pi/2, -\pi/2 + \epsilon]$ (indicated by the darkly shaded region in figure 2.11) do not get flipped around, and hence cancel each other instead of reinforcing. This results in poor estimates of the

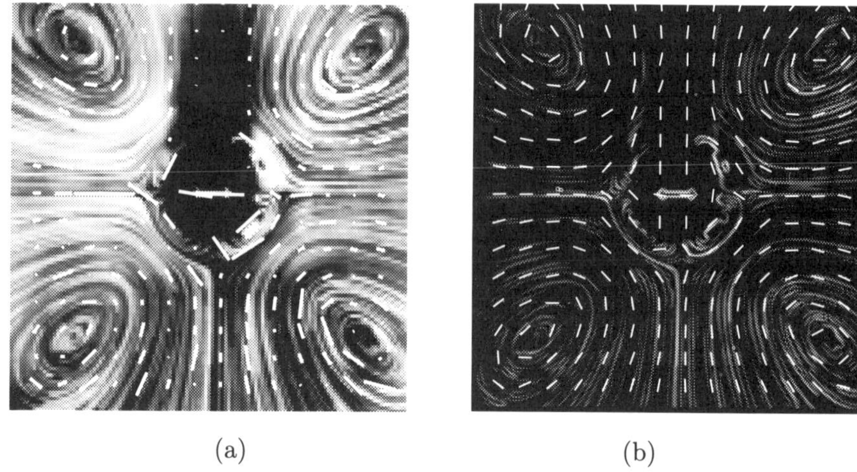

(a)   (b)

FIGURE 2.5.

(a) An image of a flow pattern induced by an oscillating cylinder *(Photograph courtesy M. Van Dyke)*. The estimated flow directions (calculated from equation 2.16) are represented by line segments overlayed on the original image. Filter sizes used were $\sigma_1 = 5$ and $\sigma_2 = 7$. The length of each line segment is proportional to the coherence at that point. Thus this image directly encodes the information about flow direction and flow coherence. (b) The coherence map. The coherence at each point of the original image (calculated from equation 2.24) is encoded as an intensity value. Filter sizes used were $\sigma_1 = 5$ and $\sigma_2 = 7$. Unit vectors representing the estimated flow directions are superimposed on the coherence map. Note that the coherence is low within the cylinder.

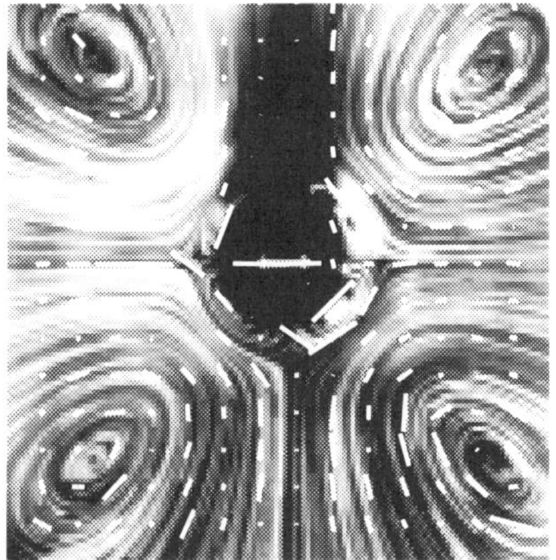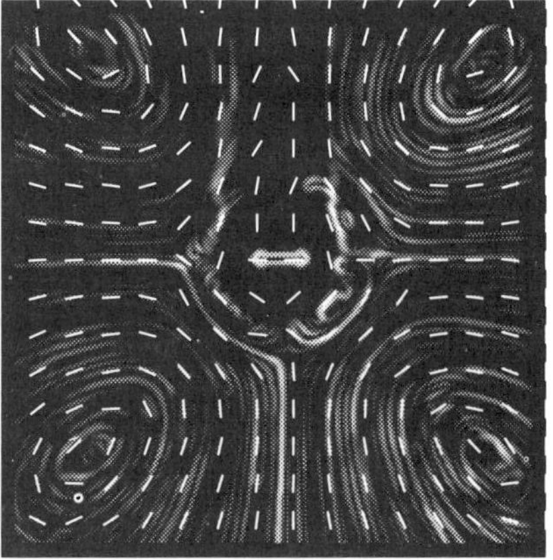

(c)

FIGURE 2.5.
(c) The estimated flow directions (calculated from equation 2.16) using filter sizes of $\sigma_1 = 9$ and $\sigma_2 = 13$. Unit vectors representing the estimated flow directions are superimposed on the coherence map.

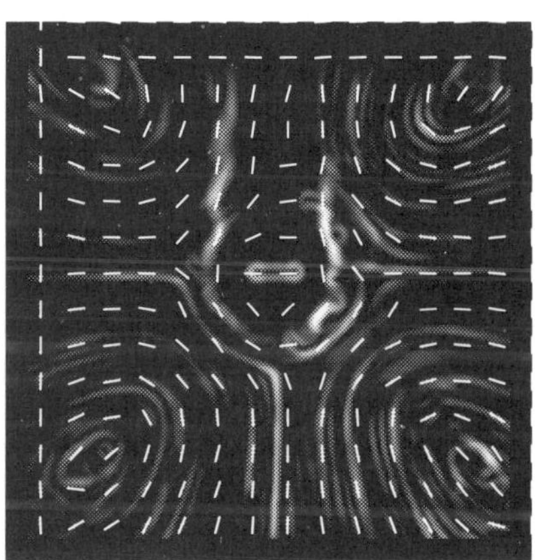

(d)

FIGURE 2.5.
(d) The estimated flow directions (calculated from equation 2.16) using filter sizes of $\sigma_1 = 15$ and $\sigma_2 = 21$. Unit vectors representing the estimated flow directions are superimposed on the coherence map. Note that the nature of figures 2.5 (b), (c) and (d) has not changed even though the scale has changed significantly.

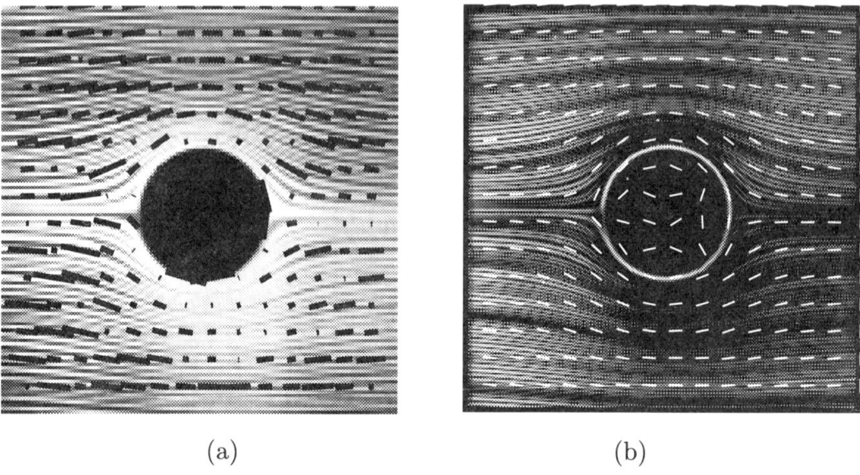

(a)   (b)

FIGURE 2.6. (a) An image of flow past a circle, with estimated flow directions overlayed *(Photograph courtesy D. H. Peregrine)*. Filter sizes used were $\sigma_1 = 5$ and $\sigma_2 = 7$. (b) The coherence map. Filter sizes used were $\sigma_1 = 5$ and $\sigma_2 = 7$.

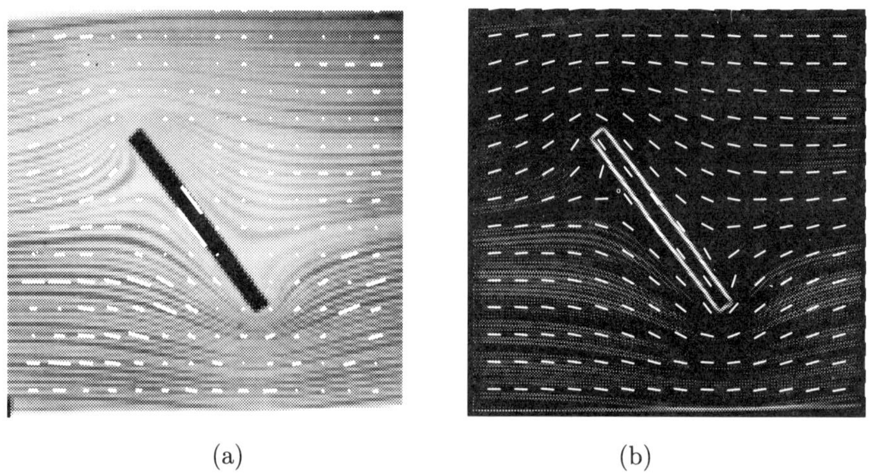

(a)   (b)

FIGURE 2.7. (a) An image of flow past an inclined plate, with estimated flow directions overlayed. Filter sizes used were $\sigma_1 = 5$ and $\sigma_2 = 7$ *(Photograph courtesy D. H. Peregrine)*. (b) The coherence map. Filter sizes used were $\sigma_1 = 5$ and $\sigma_2 = 7$.

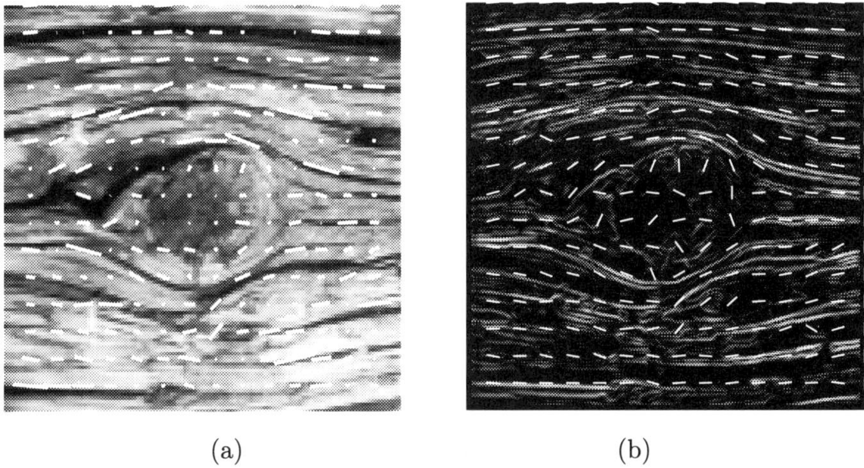

(a)   (b)

FIGURE 2.8. (a)   An image of a wood grain with a knot in the center. The estimated flow directions are overlayed on the original image. Filter sizes used were $\sigma_1 = 5$ and $\sigma_2 = 7$. (b)   The coherence map. Filter sizes used were $\sigma_1 = 5$ and $\sigma_2 = 7$.

orientation when the pattern has a horizontal flow direction.

2.5.2   COMPARING MEASURES OF COHERENCE

The result of using the coherence measure in equation 2.24 is shown in figure 2.12(a). By comparing figures 2.12 (a) and (b) one can see that equation 2.24 gives better results than equation 2.25. The coherence image corresponds closely with the original image, showing that the coherence measure that we have proposed is robust and accurate. We found this coherence measure to give better results than the measure proposed by Kass and Witkin.

2.6   Analyzing texture at different scales

There are many textures, such as the herringbone weave, which exhibit different behaviours at different scales. At finer scales, the pattern has more than one dominant orientation, but at coarser scales, the pattern has only one orientation.

This trend is clearly seen in the results shown in figure 2.13. This figure depicts a homespun woolen cloth, taken from [19]. At finer scales, the pattern appears to have alternating bands, where every other band has a texture that is diagonally oriented. At coarser scales, the pattern begins to appear vertical. This is the behaviour that one would expect from the

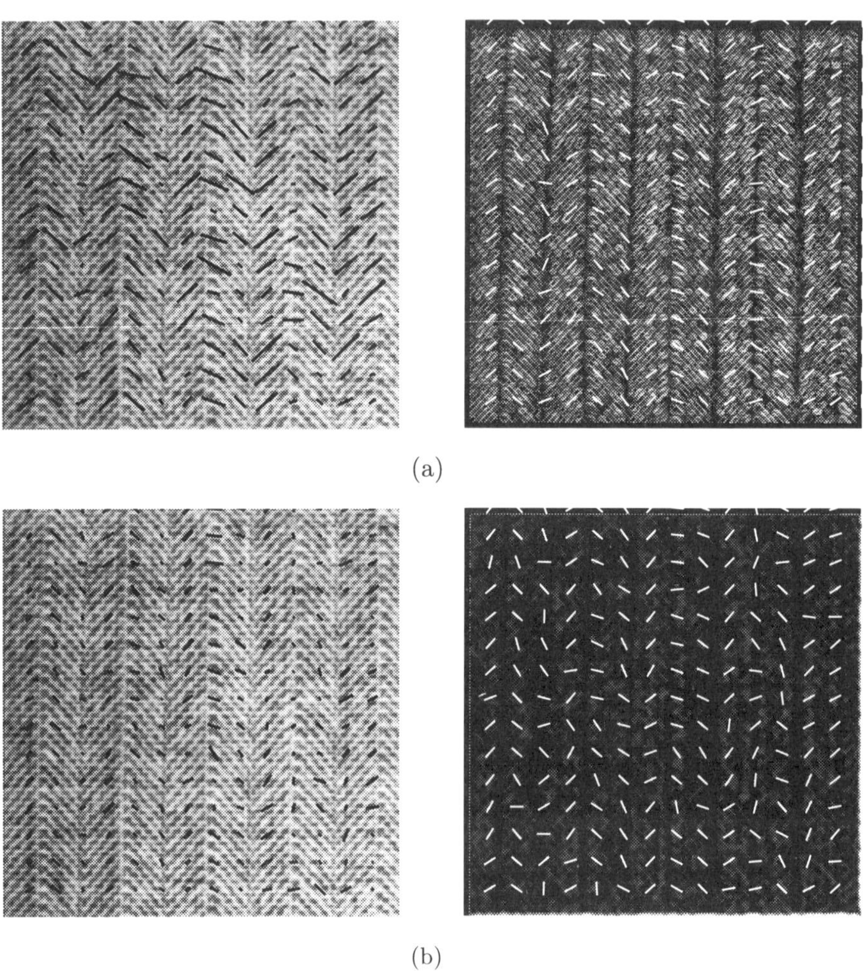

(a)

(b)

FIGURE 2.9.
Comparing the two schemes for measuring local orientation, as described
by equations 2.16 and 2.20. **(a)** shows the result obtained by using equation 2.16, and **(b)** shows the result obtained by using equation 2.20 on the
same image.

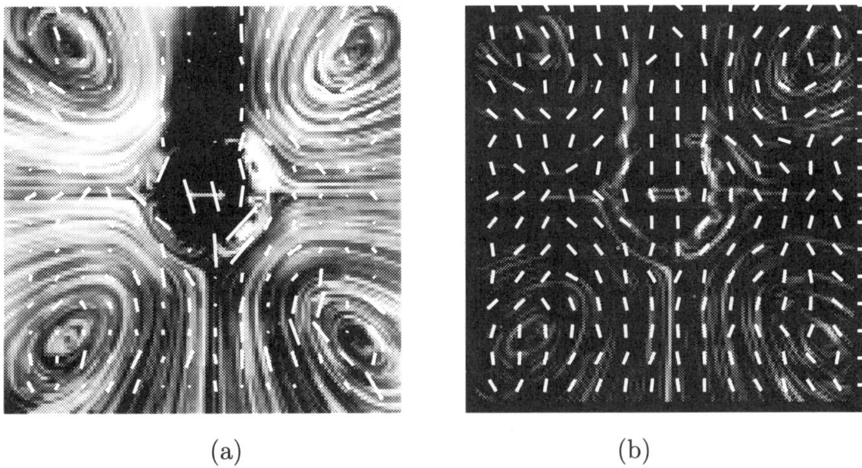

(a)   (b)

FIGURE 2.10.
Comparing two techniques for computing the angle of orientation. This figure shows results obtained by using equation 2.21. Note that the algorithm does not perform well on horizontally oriented patterns.

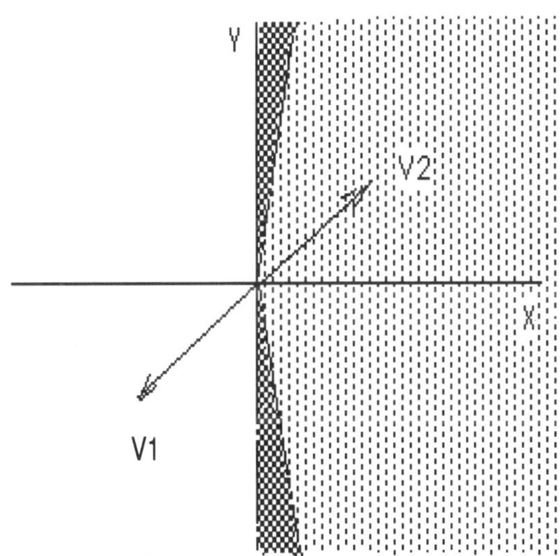

FIGURE 2.11. This figure shows how the arctangent of one argument can be used to map vectors in the xy-plane onto a half-plane. Thus, V1 gets mapped onto V2. The darkly shaded area represents those regions where vectors can point in opposite directions, and can still cancel each other out.

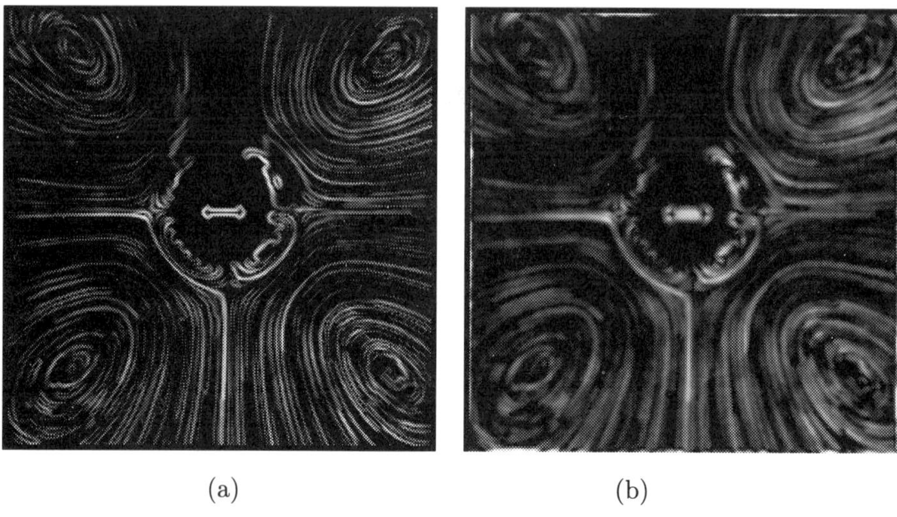

(a)   (b)

FIGURE 2.12.
(a) The coherence map obtained using equation 2.24. This gives better results than using equation 2.25, which is shown in the figure on the right.
(b) The coherence map obtained using equation 2.25

orientation estimation algorithm, and figure 2.13 proves that our algorithm indeed exhibits this behaviour.

Figure 2.13(a) shows the result of applying the orientation estimation algorithm at a scale of 5 to compute the gradient vectors, and a scale of 7 to smooth the gradient vectors. At this scale, the algorithm picks out the diagonal stripes in the texture. Figure 2.13(b) shows the result of applying the orientation estimation algorithm at a scale of 11 to compute the gradient vectors, and a scale of 17 to smooth the gradient vectors. At this scale, the diagonal nature of the alternating bands begins to weaken, and starts to tend towards the vertical at areas close to the boundaries between the bands. Figure 2.13(c) shows the result of applying the orientation estimation algorithm at a scale of 15 to compute the gradient vectors, and a scale of 21 to smooth the gradient vectors. At this scale, the texture appears predominantly vertical.

These results show that in order to analyze certain textures, one must employ different scales, because the description of the texture may vary according to the scale employed. There are some textures such as the homespun woolen cloth, and the herringbone weave [19] which appear quite different at coarse and fine scales. We propose to extend our work in this direction in the future, where we will employ a bank of filters, tuned at different scales in order perform orientation estimation. One can then use measures based on the coherence of the output to determine the scales at

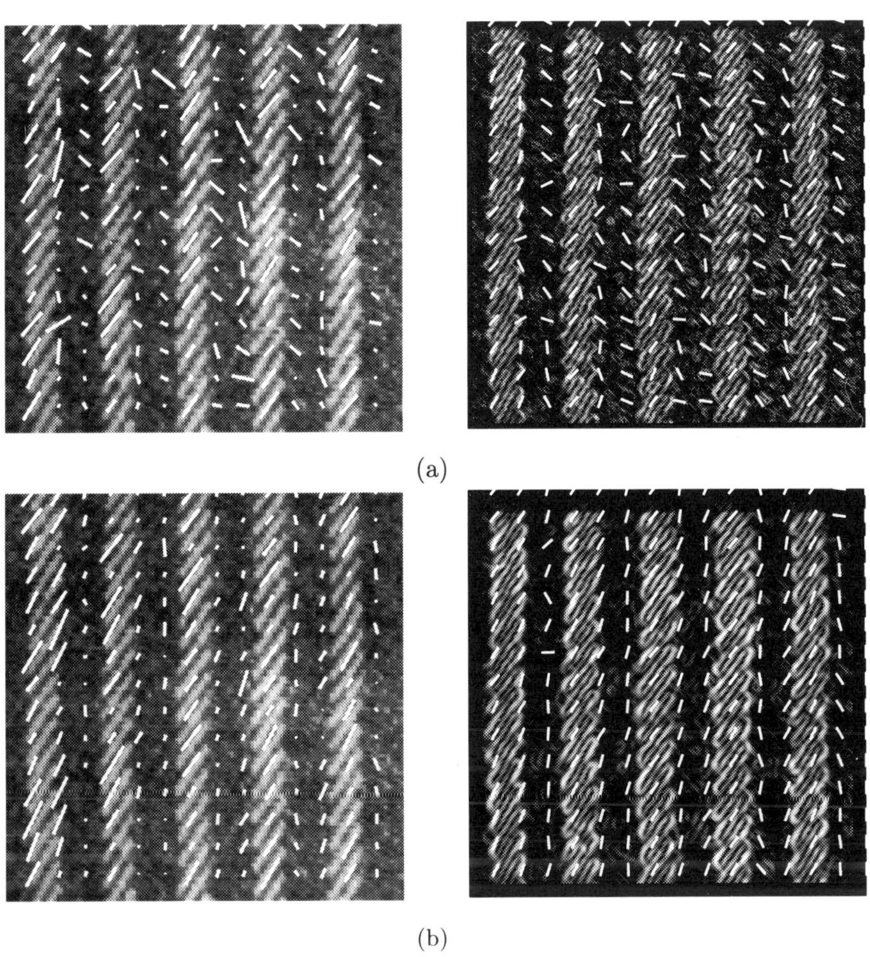

(a)

(b)

FIGURE 2.13.
(a) Analyzing the behaviour of the texture at varying scales. This figure shows the angle and coherence maps obtained by using a filter of window size 5 pixels to compute the gradient and a filter of window size 7 pixels to smooth the gradient vectors. (b) This figure shows the angle and coherence maps obtained by using a filter of window size 11 pixels to compute the gradient and a filter of window size 17 pixels to smooth the gradient vectors.

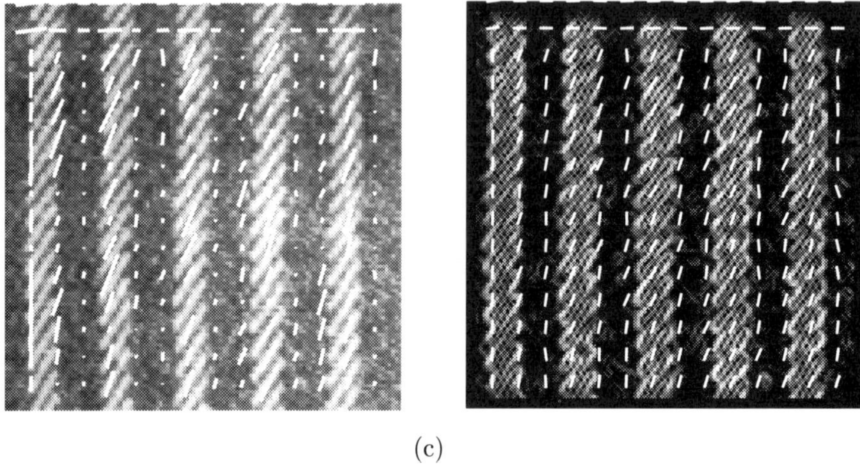

(c)

FIGURE 2.13.
(c) This figure shows the angle and coherence maps obtained by using a filter of window size 15 pixels to compute the gradient and a filter of window size 21 pixels to smooth the gradient vectors.

which the description of the texture is strong. In this manner, one can have rivalrous descriptions of textures such as the homespun cloth, where at one scale the texture appears to have certain orientations, and at another scale, the texture has different orientations.

Zucker and Cavanaugh [135] have performed experiments with subjective figures in texture discrimination. They investigate the Ehrenstein illusion, which involves two patterns – the Ehrenstein pattern, and the control pattern. The Ehrenstein pattern consists of horizontal and vertical segments. The areas bounded by the endpoints of these segments appear as disks. Figure 2.14(a) shows the underlying Ehrenstein pattern. Figure 2.14(b) shows the corresponding orientation field. Figure 2.14(c) shows the corresponding coherence image, with unit vectors overlayed. Figures 2.14(d), (e) and (f) show similar results for the control pattern, which is obtained from the Ehrenstein by shifting the horizontal segments by 0.5 cycles. Now let us rotate the Ehrenstein pattern by 45^0 about an axis perpendicular to the plane of the paper. Instead of an array of disks, subjective brightness stripes are seen, which are vertically oriented. However, the control pattern does not undergo any qualitative change as a function of orientation. Figures 2.14(g) through (l) show the orientation field for the rotated patterns. The results obtained by running the orientation estimation algorithm do not indicate the presence of subjective brightness stripes in the rotated pattern. This is because the coherence in the areas away from the line segments is very low, and hence the estimated direction of orientation within these areas has little meaning. This seems to indicate that the orientation estimation algorithm does not exhibit the subjective effects that humans do in this case. This is to be expected, because if the algorithm did exhibit the illusion, then it would not be independent of orientation of the texture, which is contrary to the fact that it is mathematically so.

2.7 Processing of the intrinsic images

In keeping with the philosophy of Barrow and Tenenbaum [8], the angle and coherence images can be regarded as intrinsic images obtained from the original oriented texture image. Intrinsic images were intended to capture a more invariant and more distinguishing description of surfaces than raw light intensities [8]. The recovery of these intrinsic images was based on domain-independent constraints generated through the physics of image formation. The first few intrinsic images proposed were that of range, orientation, reflectance and incident illumination. The orientation intrinsic image, as originally used by Barrow and Tenenbaum [8] refers to surface orientation, as described by surface normals at each point in the image. However, we use the term *texture orientation intrinsic images* to describe local orientation specificity of patterns on a two dimensional surface.

The use of the term 'texture orientation intrinsic images' is justified because the spirit in which these images are extracted and processed echoes the principles cited above. Firstly, these images are extracted by using domain-independent processing. Secondly, they are independent of lighting conditions, because of the process of averaging and normalization in computing the orientation and coherence. Thirdly, domain-dependent knowledge and constraints can be imposed on the intrinsic images to yield appropriate conclusions in specific contexts. The viewpoint we have taken is also close to that of Marr [84], who proposed the extraction of a 'primal sketch', based on domain-independent processing.

We strongly advocate the use of the angle and coherence images as intrinsic images. Almost every analysis of oriented textures will require the computation of these intrinsic images as a first step. In this sense, the computation of the orientation field, resulting in the intrinsic images is indespensible in the analysis of oriented textures. In order to justify this claim, we provide results from a number of experiments to indicate the usefulness of the angle and coherence intrinsic images.

Let us consider a specific domain in order to illustrate the application of our algorithm to real images. The automation of lumber defect detection is vital for the future control of lumber processing, which is an important industry [113]. Defects in wood are rich in textural content, and typically involve the identification of knots, worm holes, checks, and grains [25, 96, 33, 23]. As observed in [23], tonal measures alone are not sufficient to isolate these defects. For instance, both checks and splits are darker than clear wood, just as knots are.

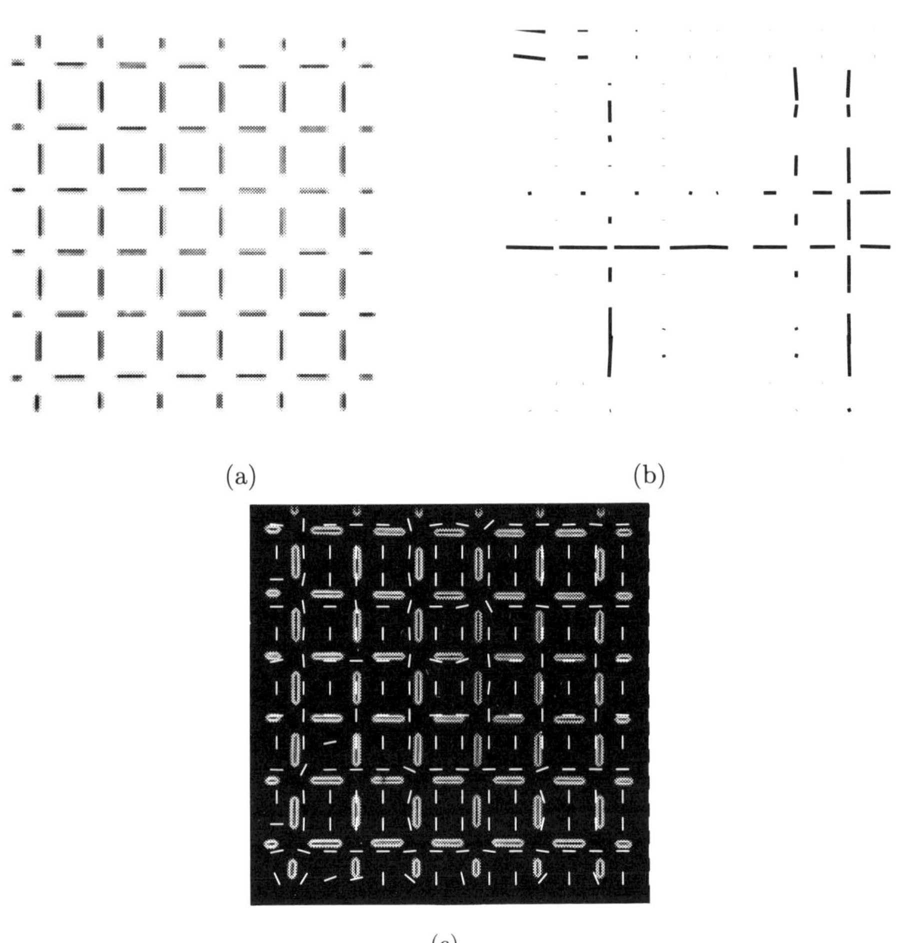

(a)

(b)

(c)

FIGURE 2.14.
(a) The original Ehrenstein pattern (b) Its orientation field. Filter sizes $\sigma_1 = 5$ and $\sigma_2 = 7$ pixels were used. (c) The coherence map

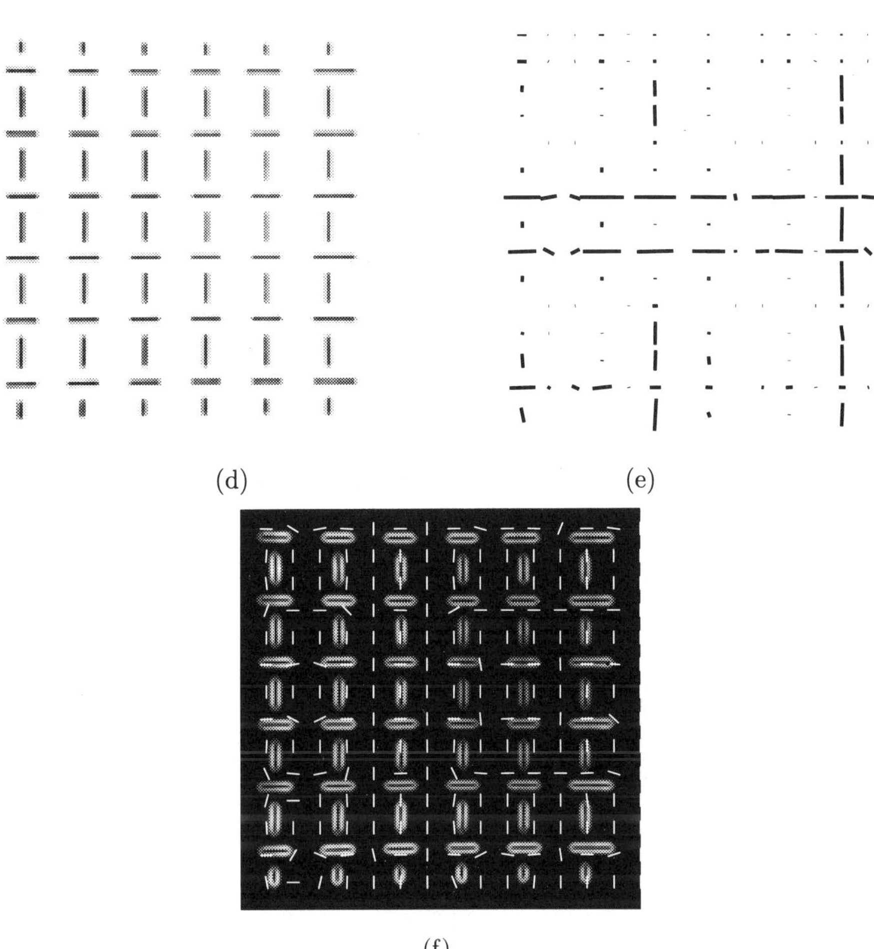

(d)

(e)

(f)

FIGURE 2.14.
(d) The control pattern and (e) its orientation field. Filter sizes $\sigma_1 = 5$ and $\sigma_2 = 7$ pixels were used. (f) The coherence map

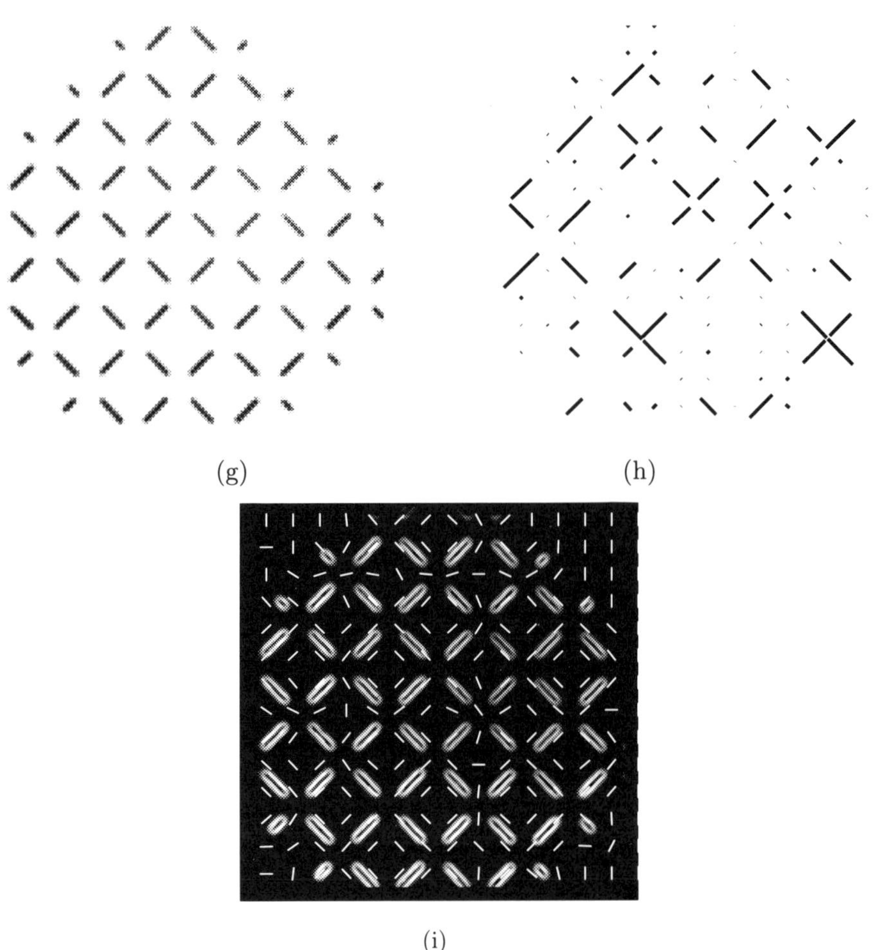

(g)

(h)

(i)

FIGURE 2.14.
(g) The Ehrenstein pattern rotated by 45^0, and (h) its orientation field.
Filter sizes $\sigma_1 = 11$ and $\sigma_2 = 17$ pixels were used. (i) The coherence map

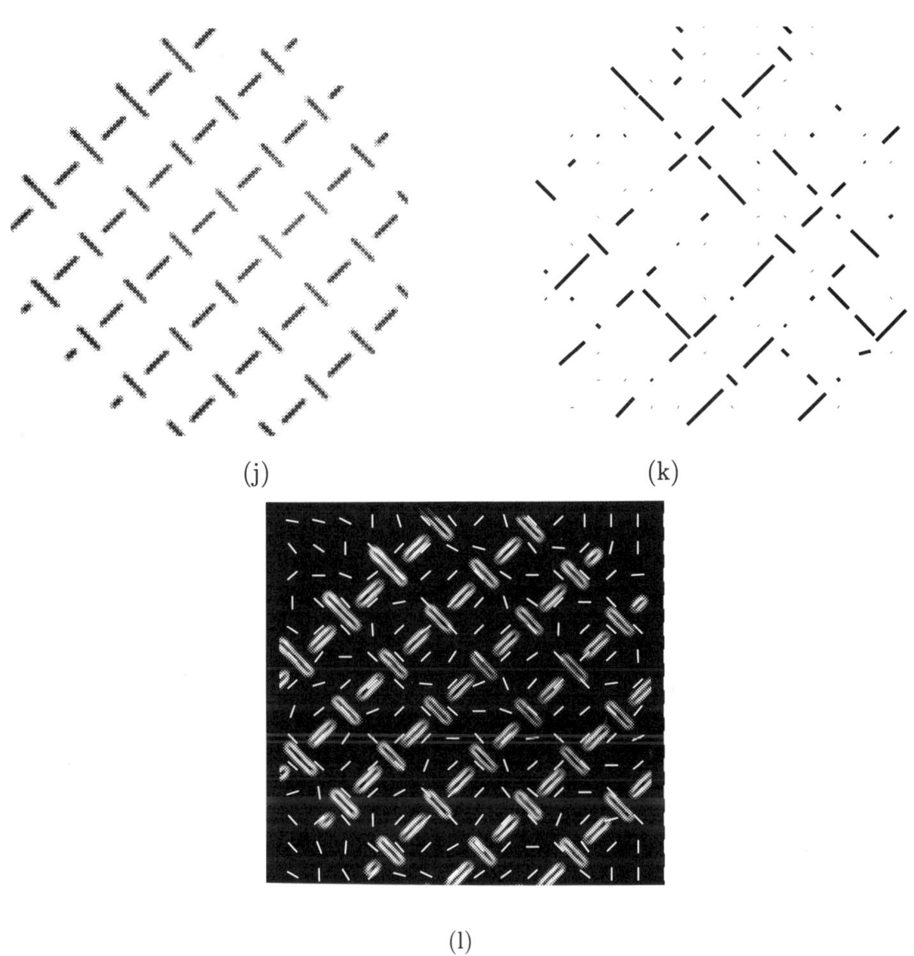

(j)   (k)

(l)

FIGURE 2.14.
(j) The rotated control pattern and (k) its orientation field. Filter sizes $\sigma_1 = 11$ and $\sigma_2 = 17$ pixels were used. (l) The coherence map

In order to illustrate the importance and usefulness of the texture orientation intrinsic images, we have used them to identify some of the defects that occur in wood, such as knots and worm holes.

2.7.1 THE COHERENCE INTRINSIC IMAGE

The coherence image ecnodes the degree of anisotropy at each point of the original image. Coherence is low within knots since there is no anisotropy of the texture within this region. Hence, the coherence image can be used to isolate knots, as these are locations in the oriented pattern that have low coherence. Figure 2.15(a) shows an image of a piece of wood containing three knots, obtained from [33, fig. 10]. Figure 2.15(b) shows the result of computing the coherence of this image, by using equation 2.22. The coherence has been encoded as a gray value, with darker regions representing low coherence values. Recall that equation 2.22 computes a coarser measure of coherence than equation 2.24, as the latter incorporates the gradient magnitude into its computation. Equation 2.24 would be suitable in an application where one needed a very detailed coherence map. However, equation 2.22 proves to be adequate in the isolation of knots, since it is a smoother measure of coherence. Figure 2.15(c) shows the result of thresholding the coherence image to produce a binary image. One can see that the locations of the knots have been isolated in the binary image, thus showing that simple processing of the coherence image yields meaningful results.

Figure 2.16(a) shows an image of a piece of wood containing worm holes, obtained from [23]. Figure 2.15(b) shows the result of computing the coherence of this image, by using equation 2.22. The positions of the worm holes correspond to regions of low coherence, ie the dark regions. Figure 2.16(c) shows the result of thresholding the coherence image to produce a binary image. One can see that the locations of the worm holes have been isolated in the binary image.

Note that the spirit of the orientation estimation algorithm is the same as that of edge detection, as the early stages of processing are identical - they both involve smoothing and the computation of the gradient. However, the post processing reflects different goals. The goal of edge detection is to strictly *localize* the position of sharp gradients, while the goal of orientation estimation is to *combine* gradient information within a neighborhood to infer a dominant local orientation.

The coherence image can be processed in a variety of ways. Figure 2.17(a) shows the image of a brick wall, obtained from [19]. The coherence measure in equation 2.24 is used to generate the coherence image of figure 2.17(b). The coherence image is then thresholded with a *low* threshold, so as to isolate regions having low coherence (the brighter a region, the higher its coherence). The resulting image is shown in figure 2.17(c). Note that the interior of the bricks are regions with low coherence, and are hence isolated

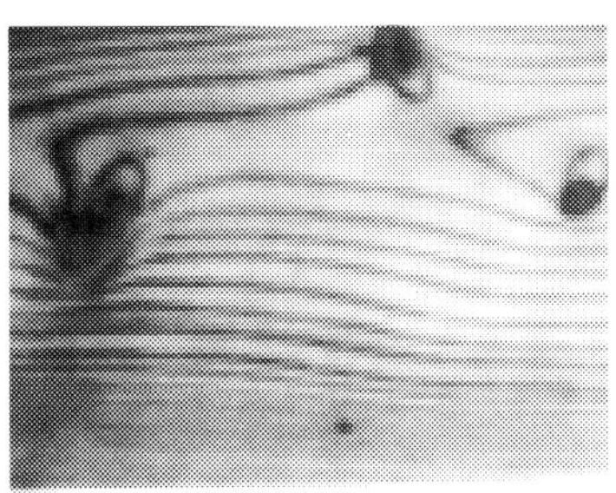

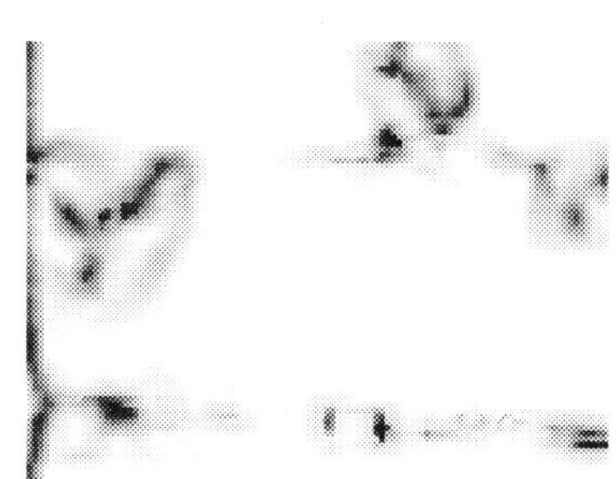

(a)   (b)

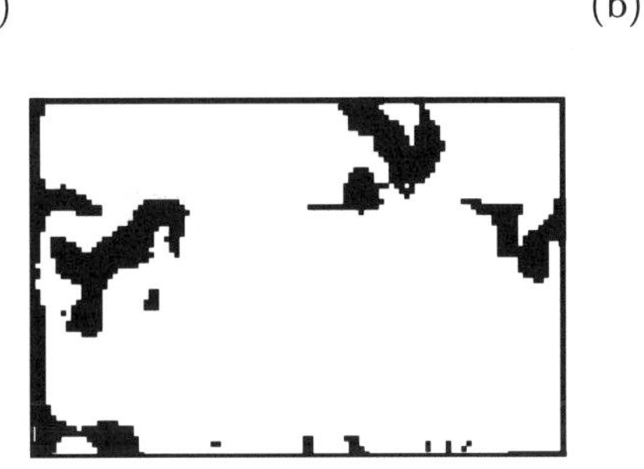

(c)

FIGURE 2.15.
(a): An image of a piece of wood containing three knots (Reproduced with permission from *Defects in Wood*, by W. Erteld, W. H. Metta and W. Ac-terberg, 1964 (Leonard Hill) ©Blackie and Son Ltd., Glasgow and London). (b): The coherence image calculated using filter sizes $\sigma_1 = 5$ and $\sigma_2 = 7$ pixels ; (c): The thresholded coherence image. Note that the locations of the knots show up in the thresholded image

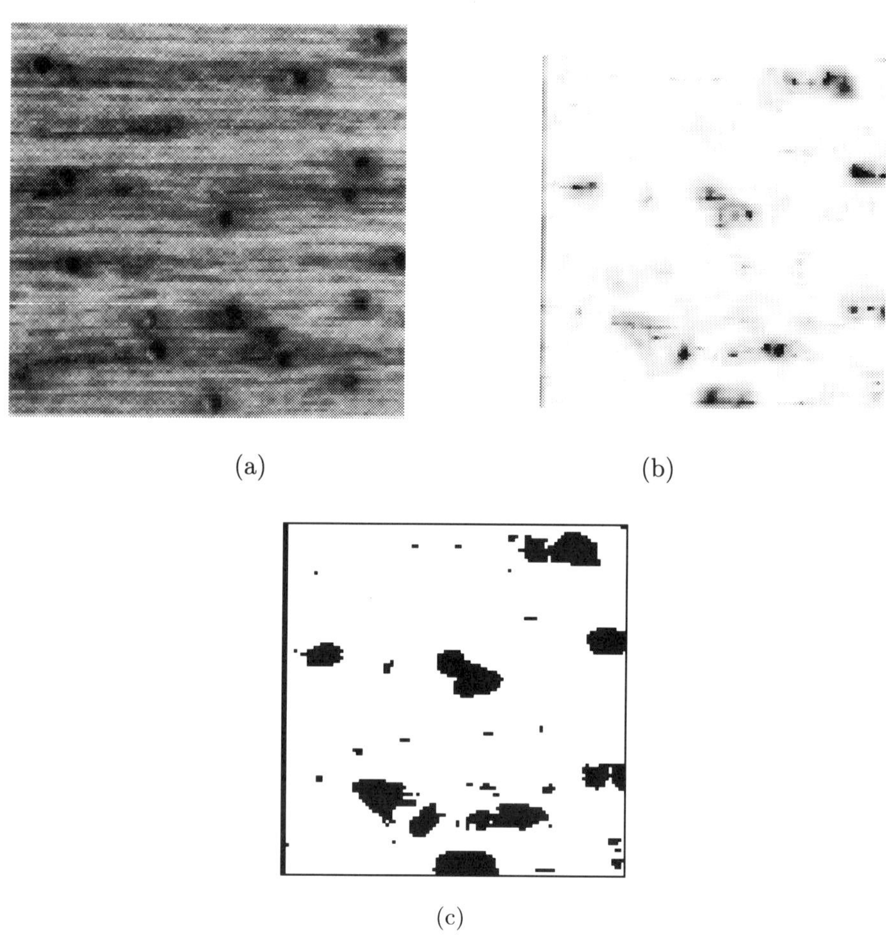

(a)

(b)

(c)

FIGURE 2.16.
(a): An image of a piece of wood containing worm holes (From Conners
et al [23], ©1983 IEEE). (b): The coherence image, calculated using filter
sizes $\sigma_1 = 5$ and $\sigma_2 = 7$ pixels ; (c): The thresholded coherence image. Note
that the locations of the worm holes show up in the thresholded image

in the thresholded image. This can then be used to locate the positions of individual bricks within the lattice.

Another class of defects that appears in wood is that of *checks* [96, p284]. Checks are longitudinal openings at weak points in the wood. One way of determining the gross 'checkiness' of a piece of wood is to simply compute its average coherence. The larger the number of checks, or the more severe the intensity of the checks, the larger the coherence measure will be. Hence a simple averaging of the coherence image suffices to obtain a global measure of the number of checks on the wood.

In section 1.4.2, chapter 1 we considered the description of an image showing an orange peel defect. An *orange peel* defect arises when the surface texture appears wrinkled, like the skin of an orange. Such a wrinkled texture is actually an oriented texture, and the method described in this section can be used.

Figure 2.18(a) shows an image of an orange peel defect. The sample consists of a silicon wafer that has undergone an isotropic etch using hydrofluoric and nitric acids. The surface has a wrinkled appearance, and the appropriate classification for this type of defect is to term it a weakly ordered texture. Figure 2.18(b) shows the orientation field overlayed on the original image. The orientation field is portrayed by white segments, such that the direction of the segment corresponds to the orientation of the underlying texture, and the length of the segment is proportional to the coherence of the texture.

Currently, inspection technicians use terms like 'light' orange peel and 'heavy' orange peel to describe the severity of the defect. However, no precise quantitative measure is available. Hence, we propose the following measure for the severity of an oriented texture defect, such as orange peel. Find the average coherence $\bar{\rho}$ over the image of the defect, where ρ is defined in 2.24.

In the case of the image shown in figure 2.18, the average coherence measure for the region of orange peel is 0.13, which represents the severity of the defect. Of course, other measures for the amount of orange peel can also be defined, but we have given one possible measure in this chapter.

Figure 2.19(a) shows a similar type of surface defect on a GaAs wafer. The image depicts an etched line, which consists of photoresist deposited on a GaAs substrate. The photoresist exhibits ripples or undulations, and the appropriate classification for this type of defect is to term it a weakly ordered texture. Figure 2.19(b) shows the orientation field overlayed on the original image. In the case of the image shown in figure 2.19, the average coherence, $\bar{\rho}$, over the rippled surface is 0.28 (where ρ is defined in 2.24). which represents the severity of the defect.

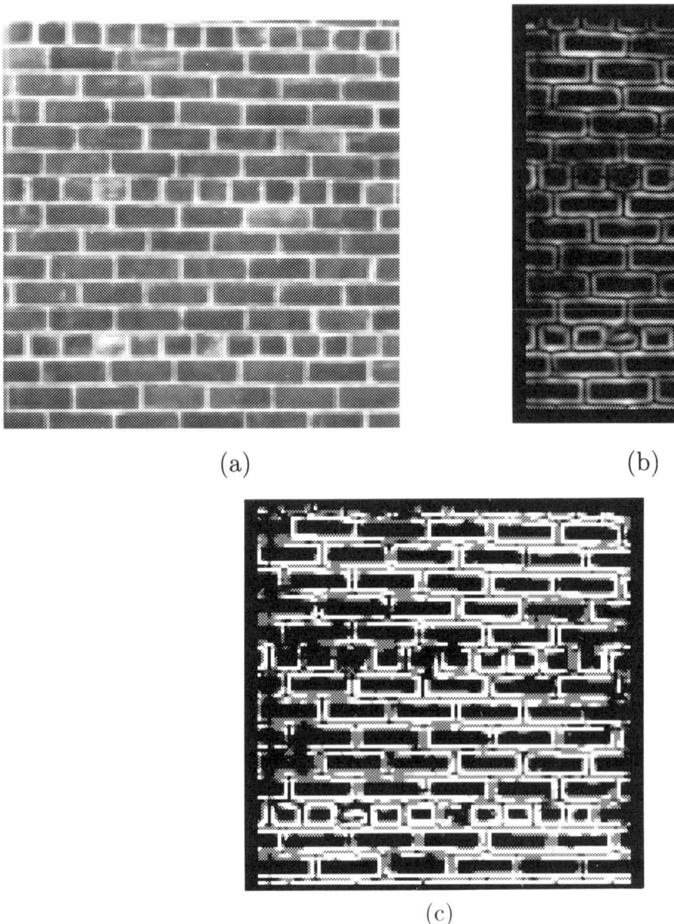

(a)

(b)

(c)

FIGURE 2.17.
(a): An image of a brick wall; (b): The coherence image, calculated using
filter sizes $\sigma_1 = 9$ and $\sigma_2 = 13$ pixels ; (c): The thresholded coherence
image. Note that the interior of the bricks show up in the thresholded
image

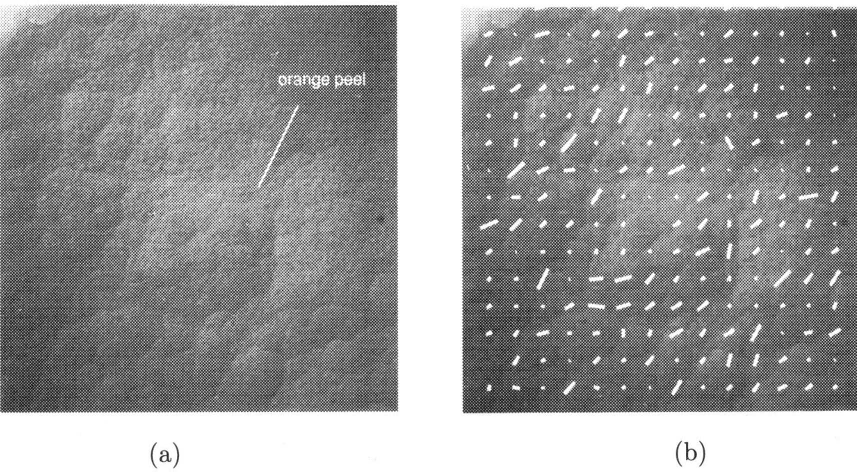

(a)   (b)

FIGURE 2.18. (a): The original image of the orange peel defect, obtained at a magnification of 500, using Nomarski phase contrast (b) The orientation field overlayed on the original image in the form of white segments.

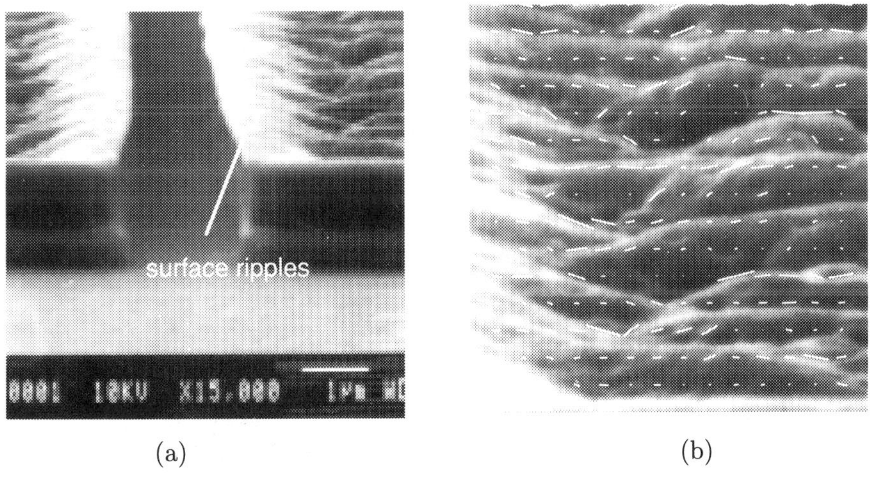

(a)   (b)

FIGURE 2.19. (a): An image of an etched line showing surface ripples (b) An enlarged portion of the surface ripple region, showing the overlayed orientation field

2.7.2 THE ANGLE INTRINSIC IMAGE

When the fiber alignment in a piece of wood does not coincide with the longitudinal axis of the piece, the wood is said to be *cross-grained* [96, p. 251]. Any form of cross grain which occurs in structural lumber is a defect because of the reduction in the strength of the member in which it occurs. The direction of grain may be determined by observing the orientation of resin canals or vessels.

The angle intrinsic image readily yields this information. All one needs to do is histogram the angle image, and locate the peak of the histogram. This will yield the most dominant global orientation in the texture, and will coincide with the direction of the grain. In order to estimate the dominant global directions, one can compute an angle histogram, weighted according to the coherence measure. Let us map all the angles into the range $[0, \pi)$, and discretize this range into 180 bins, each bin corresponding to the angle in degrees. Let the angle at point (i, j) be θ_{ij} degrees, where θ_{ij} is quantized to be an integer in the range $[0,180)$. Let the coherence at this point be ρ_{ij}. Let H denote the histogram, and H_k be the kth bin of the histogram. In order to compute the weighted histogram, the contribution of point (i, j) is then ρ_{ij} towards bin θ_{ij}. These contributions are summed over all image points, in order to yield an angle histogram. Thus,

$$H_k = \sum_{i,j=1,N} (\rho_{ij} \mid \theta_{ij} = k) \qquad (2.26)$$

Figure 2.20(a) shows an image of a piece of wood containing checks, obtained from [70][p. 91]. In order to find out whether there is a dominant orientation, one needs to compute the orientation field, which is shown in Figure 2.20(b). Figure 2.20(c) shows the weighted angle histogram, computed according to equation 2.26. One can see that there is a very sharp peak at $\theta = 140^0$, showing that the pattern is strongly unidirectional, and the dominant direction is 140^0. Thus, one can easily obtain reliable estimates of dominant orientation in such cases. (In all the figures displayed, the angle is measured clockwise with respect to the x axis, which points horizontally to the right).

Figures 2.21 and 2.22 show another application of the angle histogramming technique, where the dominant orientation is computed over different scales. As mentioned in section 2.6, the description of a texture may change over different scales. One way this can happen is that the dominant orientation of a texture can change dramatically when observed at different scales. Figure 2.21(a) shows an image of a herringbone weave, digitized to a size of 240x240 pixels, obtained from [19]. Figure 2.21(b) shows the result of applying the orientation estimation algorithm at a scale of 5 to compute the gradient vectors, and a scale of 7 to smooth the gradient vectors. At this scale, the texture has two distinct directions, as shown by the weighted angle histogram in figure 2.21(c). There are strong peaks at 40^0 and 140^0.

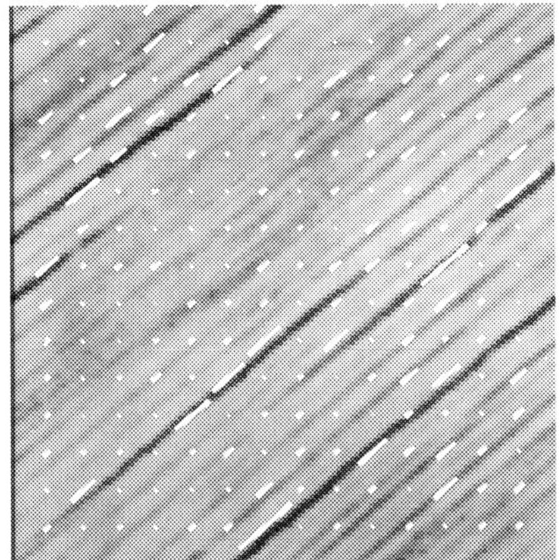
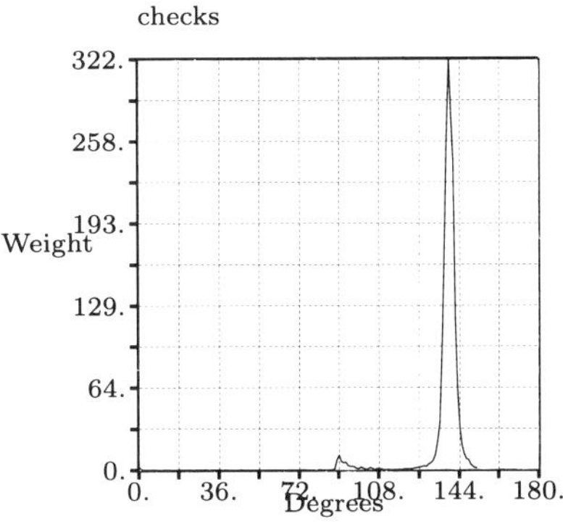

FIGURE 2.20.
(a):An image of a piece of wood with checks and its orientation field computed for filter sizes $\sigma_1 = 5$ and $\sigma_2 = 7$ pixels (b): The weighted angle histogram, showing a sharp peak at 140^0, indicating that the texture is strongly unidirectional, with an orientation of $\theta = 140^0$.

A dramatically different picture emerges when the orientation estimation algorithm is applied at a scale of 15 to compute the gradient vectors, and a scale of 21 to smooth the gradient vectors. Figure 2.21(d) shows the result of applying the orientation estimation algorithm, and figure 2.21(e) shows the weighted angle histogram. Figure 2.21(e) displays a completely different behaviour from figure 2.21(c), because there is only one sharp peak in the histogram at 90^0. This shows that the texture appears predominantly vertical at higher scales, and distinctly bidirectional at lower scales.

Figure 2.22 illustrates a similar behaviour in the case of an image of homespun cloth, obtained from [19]. Figure 2.21(b) shows the result of applying the orientation estimation algorithm at a scale of 5 to compute the gradient vectors, and a scale of 7 to smooth the gradient vectors. At this scale, the texture has one dominant direction, at 120^0, as shown by the weighted angle histogram in figure 2.22(c). Figure 2.21(d) shows the result of applying the orientation estimation algorithm at a scale of 15 to compute the gradient vectors, and a scale of 21 to smooth the gradient vectors. At this scale, the texture has two dominant directions, at 120^0 and 90^0, as shown by the weighted angle histogram in figure 2.22(e). Thus, one can see that the description of oriented textures is dependent on the notion of scale,

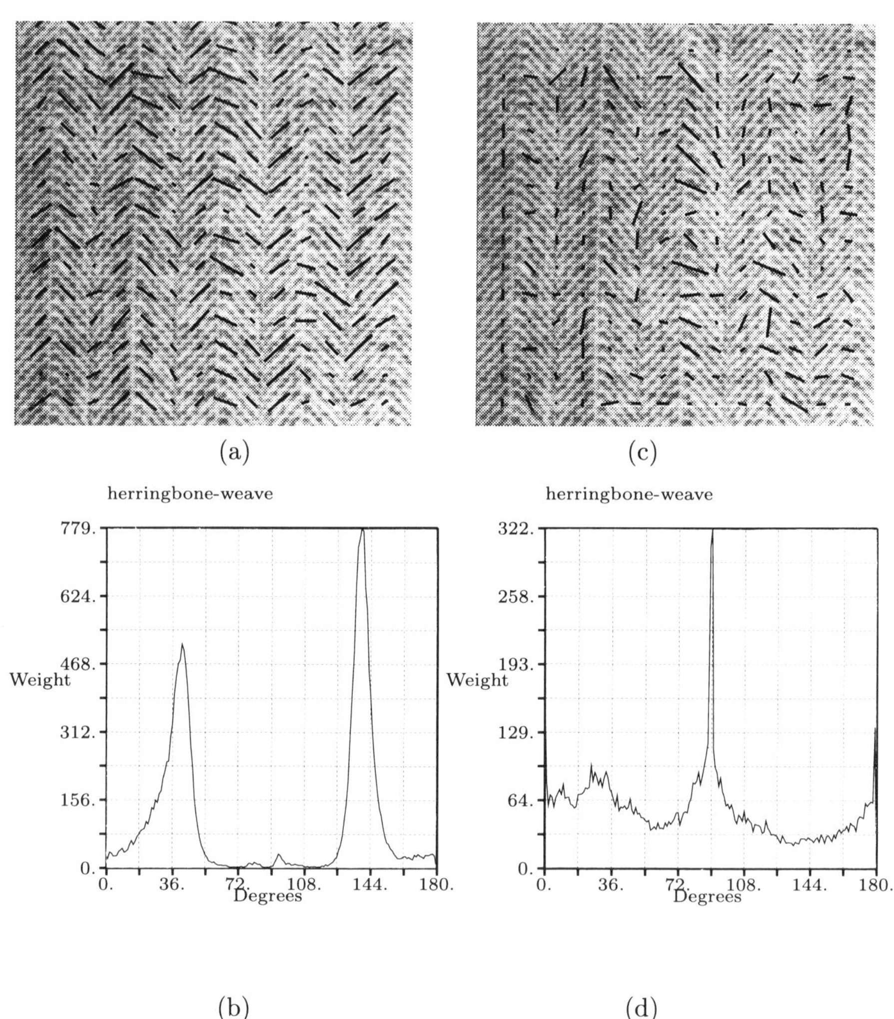

FIGURE 2.21.
(a) An image of a herringbone weave, from [19] with its orientation field computed for filter sizes $\sigma_1 = 5$ and $\sigma_2 = 7$ pixels (b): The weighted angle histogram, showing sharp peaks at 40^0 and 140^0. (c): The orientation field computed for filter sizes $\sigma_1 = 15$ and $\sigma_2 = 21$ pixels (d): The weighted angle histogram, showing a sharp peak at 90^0, indicating that the texture is predominantly vertical

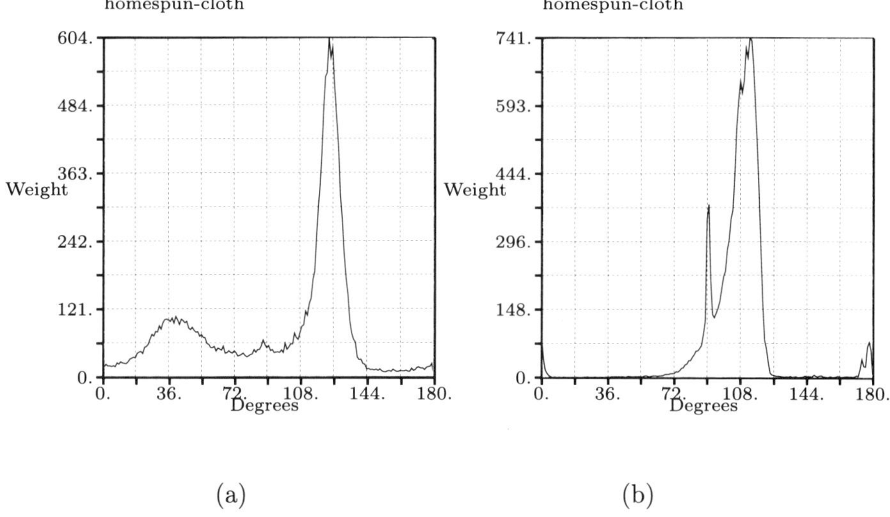

(a)   (b)

FIGURE 2.22.

This figure shows the angle histograms for the image of homespun cloth, shown in figure 2.13. The orientation fields have been displayed in figures 2.13(a) and 2.13(c) **(a)**: The weighted angle histogram, showing a sharp peak at 120^0. This was calculated from the orientation field. Filter sizes $\sigma_1 = 5$ and $\sigma_2 = 7$ pixels were used.

(b): The weighted angle histogram, showing sharp peaks at 90^0 and 120^0. This indicates that the texture has started acquiring a strong vertical component at this scale. The orientation field was computed for filter sizes $\sigma_1 = 15$ and $\sigma_2 = 21$ pixels.

2.8   Conclusions

In this chapter we presented a new algorithm for computing the orientation field for flow-like textures. The basic idea behind the algorithm is to use an oriented filter, namely the gradient of Gaussian, and perform manipulations on the resulting gradient vector field. We have proved the optimality of this algorithm in estimating the orientation direction. An added strength of the algorithm is that it is simpler and has a better signal to noise ratio than previous approaches, because it employs fewer derivative operations. We have also proposed a new measure of coherence, which works better than previous measures.

We are convinced that the angle and coherence intrinsic images represent a necessary initial stage in the processing of oriented textures. We justified the role that the texture orientation intrinsic images play by providing results from a number of experiments. These results indicate that operations on the angle and coherence images yield useful results in many domains. Given this important role that the texture orientation intrinsic images play, our results suggest the best way to compute these images.

3

The analysis of oriented textures through phase portraits

3.1 Introduction

One of the central issues in computer vision is the problem of signal to symbol transformation [44, 102], which really has two components. The first is the choice of symbols to be derived. This is usually ignored or taken for granted. The second is the exact method used to transform the given image into this set of symbols. Researchers in computer vision have usually focussed on this part of the problem. In this chapter we analyze the symbolic description of oriented textures by first proposing an appropriate set of symbols and then solving the signal to symbol transformation problem.

3.1.1 OVERVIEW OF APPROACH

Our approach to the problem of oriented texture description is to view it as a two stage process. The first stage is concerned with extracting an orientation field from the raw image. This gives rise to the angle and coherence intrinsic images. We use the algorithm developed by Rao and Schunck [106] and presented in chapter 2 in order to extract the orientation field.

The second stage is concerned with performing computations on these intrinsic images in order to derive a qualitative description of the orientation field, and hence the underlying texture. This chapter presents a method to analyze a given orientation field based on the geometric theory of differential equations [76, 7, 51, 6]. This theory involves the notion of a *phase portrait*, which represents the solution of a system of differential equations in a qualitative fashion.[1] For a linear 2D system, only a finite number of qualitatively different phase portraits are possible. This allows a parametric representation of different types of phase portraits, and their classification is possible based on the notion of qualitative equivalence. The basic idea is to view a given texture flow pattern as being comprised of piecewise linear flows, and to describe each linear flow by means of an equivalent phase portrait.

In section 3.4 we establish a connection between real images and phase

---

[1]Section 3.3 defines these terms.

portraits. This is done by proposing a distance measure between a local orientation field, derived from a local area of the given texture and a hypothesized phase portrait. Consequently, we search for a solution that minimizes this distance, i.e. the phase portrait which best matches the given orientation field. From this perspective, the problem is a non-linear least squares optimization, which we have solved using the Levenberg-Marquardt technique.

The main result of this chapter is that the theory of differential equations can be successfully used to describe flow-like textures both qualitatively and quantitatively. An attractive feature of this method is that it is completely domain independent. The only assumption make is that the texture is flow-like. We present the results of applying the method to several real texture images.

There are several advantages to using such a scheme. Firstly, this scheme offers a novel symbol system to describe oriented textures. There are many complex textures that occur in nature, and it is of interest to derive a nomenclature that humans can use in the description of such textures. Secondly, this work shows that it is possible to automatically synthesize descriptions of complex oriented patterns. Since our method is local in nature, (ie descriptions are derived over local neighborhoods), the oriented pattern can vary across the entire image. The implication of this fact is that a variety of textures can be described under the same computational paradigm, even though the textures can vary considerably. Interestingly, even for textures that are difficult to describe by humans, the program is able to pinpoint the salient features of the texture. Thirdly, the proposed scheme can be used for making quantitative measurements in the area of experimental fluid mechanics and flow visualization, as described in section 3.5.5. Finally, this scheme can be used for the segmentation of oriented texture patterns. This segmentation is rooted in the appearance of phase portraits, and is especially useful in analyzing complex images generated in experimental fluid mechanics.

3.2   Background

In this section, we address the issue of why a qualitative scheme to describe textures is both necessary and important. We also review briefly a technique to extract the orientation field for a texture, as described in [106].

3.2.1   RELATED RESEARCH

One important class of textures that has received attention recently is flow-like texture [65, 106], which is the focus of this chapter.

The research presented in this chapter is also part of a larger effort towards texture interpretation [104]. One way to systematize the interpretation and description of textures is to devise a taxonomy for texture as done by Rao and Jain [104]. Flow-like textures, which form an important part of the above taxonomy are analyzed in this chapter.

Optical flow is an important research area computer vision, and can yield information about the motion and structure of objects [99, 63]. Optical flow fields are essentially vector fields, and can be estimated by algorithms such as [49]. The orientation field for a flow-like texture bears a close resemblance to vector fields, which have been studied by researchers in motion analysis. Koenderink and van Doorn [69] decompose the motion parallax field into three elementary fields, that of a center, a node and a saddle. They classify motion parallax fields into four types, viz. node, saddle-point, focus and center, and mention that this can be used to characterize the motion parallax field in a qualitative, topological manner.

Recently, Verri and Poggio [119] have argued that meaningful information about the 3-D velocity field can be obtained from the qualitative properties of the motion field. They have used the qualitative theory of planar differential equations as a vehicle to describe qualitative properties of the motion field. They advocate the view that an optical flow can be thought of as *close* to the true motion field if the topology of the two vector fields is the same according to the qualitative theory of planar differential equations. However, no algorithm has been proposed for how a qualitative description can be derived from the motion field or the optical flow.

Another piece of interesting work which uses qualitative terms to describe a velocity field is [98]. The derivation of qualitative terms is based on the analysis of streamlines in fluid flow, and in section 3.7, we show this to be equivalent to the analysis of linear phase portraits.

Helman and Hesselink [50] have developed a representation for vector fields that involves the topology of fixed points and their connections. They report that the extraction and use of such features greatly reduces the computational burden involved in visualizing large vector fields. It is also necessary to extract qualitative information in order to visualize and interpret such complex data sets.

Yip [130] has developed a program called KAM, which qualitatively analyzes the behaviour of a dynamical system. An interesting aspect of this program is that it converts a symbolic description of a dynamical system into a geometric description by plotting discrete points on phase portraits of the system. The program generates a qualitative description of these points by aggregating them into clusters and detecting features of the clusters (e.g. whether the cluster forms a closed curve).

The underlying philosophy of this work is to determine a symbol system for oriented textures, and to describe a given texture in terms of these symbols. This agrees with the work done in 3D object recognition [15, 14, 60], where the goal has been to define primitive surface types and then to segment a given image into these types. In principle, the idea of a *symbolic surface descriptor* [60], which is a set of polynomial coefficients describing a smooth surface patch, is similar to the scheme of finding the describing coefficients of a linear phase portrait, as presented in this chapter. With this spirit, our approach could be labeled as *symbolic descriptors for oriented textures.*

3.2.2 POTENTIAL APPLICATIONS

The techniques developed in this chapter are very useful in the characterization of flow visualization pictures, as discussed in section 1.3.

Subsurface microstructure, consisting of largely flow-like textures plays a crucial role in the analysis of fractograph specimens [137]. A qualitative scheme to describe these textures would be useful for the purposes of identification. For instance, in petrography, orientation analysis is helpful in labeling certain textures as lamellar (plate-like) [78].

As discussed in section 1.3, the methods in this chapter can be used to characterize defects in wood, such as knots, and for the description of fingerprints.

The usefulness of the symbolic descriptors in this chapter has been demonstrated by applying it to defect description and classification in semiconductor wafer inspection [103].

3.2.3 OBTAINING THE INTRINSIC IMAGES

The algorithm for estimating the orientation of a texture field is based on the gradient of a Gaussian, and has been described in detail in chapter 2. Thus, one can obtain a description of the texture by using equations 2.16 and 2.24.

3.2.4 PROBLEMS IN POST-PROCESSING

One should be aware of a major limitation in using the orientation field for post-processing. The orientation estimation algorithm gives rise to an *orientation* field and **not** a vector field. The orientation estimate is inherently ambiguous, because the texture flow at any given point could be either in direction θ or $\theta + \pi$. Hence, any approach that requires as input a correct vector field will fail to yield the desired results.

Figure 3.1 illustrates this problem. Figure 3.1(a) shows a flow image [118], which displays four vortices at the four corners of the image. Figure 3.1(b) shows the result of applying the orientation estimation algorithm to this image. In this figure, directions shown by the arrowheads represent the orientation estimates from the algorithm. One can clearly see that there are many places in the orientation field where the direction of flow is not correct. This conclusion is made based on the physics of the underlying fluid flow (since the flow is laminar, there cannot be abrupt discontinuities in the velocity field). Figure 3.1(c) shows the actual direction of flow based on physical considerations such as laminarity of the flow (there is the possibility of a global sign reversal on all the arrows, but we are concerned with locally incompatible directions).

3.3   Geometric theory of differential equations

The question that we address in this section is: *given the orientation field, what higher level (symbolic) descriptors of the field are meaningful?* This will result in the compression of data as well as provide qualitative descriptions of the texture. As described in the introduction, the creation of such qualitative descriptors for textures constitutes an important task. In this section we describe how the problem is formulated with the help of concepts from the geometric approach to differential equations. The basic idea is to view a given texture flow pattern as being comprised of piecewise linear flows , and to describe each linear flow by means of an equivalent phase portrait. A number of excellent references are available on the geometric theory of differential equations, such as [6, 7, 51, 76]. The terms 'qualitative approach' and 'geometric approach' are used interchangeably in the literature, and we shall not make a distinction between them.

3.3.1   PRELIMINARY DEFINITIONS

In this section we summarize the notion of the qualitative or geometric behaviour of differential equations, as presented in [7].

3.3.1.1   One dimensional case

Consider the differential equation

$$\dot{x}(t) = \frac{dx}{dt} = X(x) \tag{3.1}$$

where X is a function. The traditional way of graphically describing the solution to equation 3.1 is to plot it in the $x - t$ plane. Let us consider instead a representation on the $x-$axis. If $X(x) \neq 0$, then the solution is either increasing or decreasing. Suppose $X(x) \neq 0$ for $x \in (a, b)$. Then the interval is labeled with an arrow showing the sense in which x is changing. When $X(c) = 0$, the point c is known as a *fixed point* [7], because x remains at c for all t. [2] A geometric representation of the qualitative behaviour of $\dot{x} = X(x)$ is called its *phase portrait*. For example, consider the equation

$$\dot{x} = x \tag{3.2}$$

---

[2] Alternate terms for a *fixed point* include a *singular point* [6], or an *equilibrium point* [54].

FIGURE 3.1. (a) Shows the original flow image.

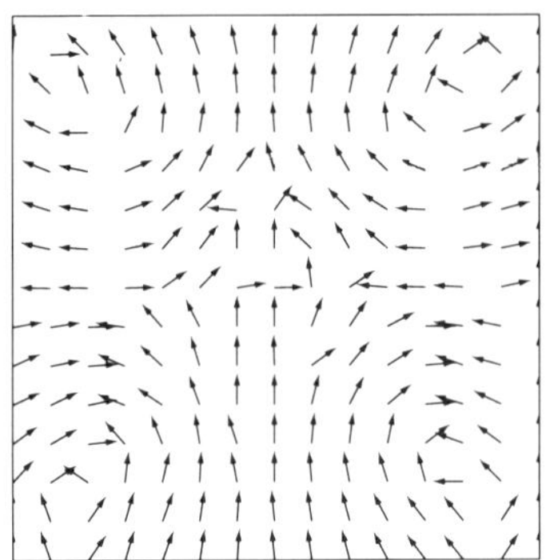

FIGURE 3.1. (b) Shows the result of applying the orientation estimation algorithm. Directions of flow are shown by the arrowheads.

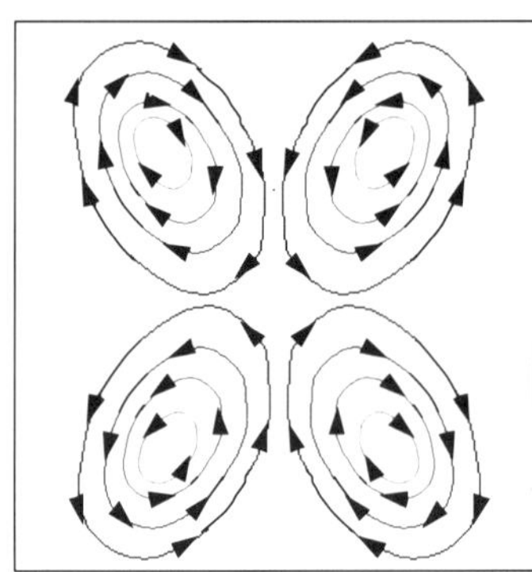

FIGURE 3.1. (c) Shows the *actual* flow directions

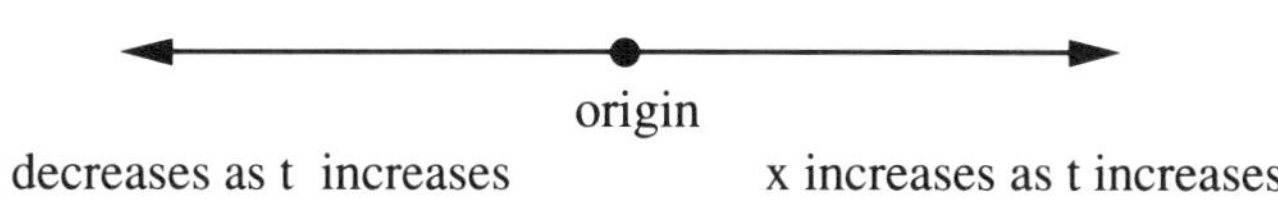

FIGURE 3.2. Illustrating the phase portrait for equation 3.2

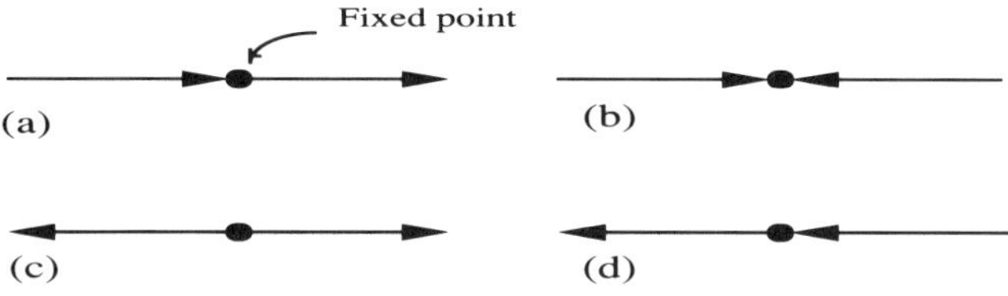

FIGURE 3.3. Illustrating the four possible phase portrait for a single fixed point

In this case $x = 0$ is a fixed point, and in the interval $(0, \infty)$ x increases as t increases, and in the interval $(0, -\infty)$ x decreases as t increases. Thus, the phase portrait for equation 3.2 looks as in figure 3.2.

Note that when x is not a fixed point, it must be either increasing or decreasing. Hence for any finite number of fixed points, there is a finite number of *different* phase portraits. The term *different* refers to distinct assignments of where x is increasing or decreasing. For instance, with one fixed point, say $x = c$, there are exactly four possible distinct phase portraits, as shown in figure 3.3.

This means that the qualitative behaviour of *any* differential equation of the form in equation 3.1 with one fixed point *must* correspond to some phase portrait in figure 3.3 for some value of c. For instance, the qualitative behaviour of $\dot{x} = x$ and $\dot{x} = x^3$ are identical, and correspond to the phase portrait in figure 3.2. This leads to the notion of *qualitative equivalence* [7], defined as follows. Two differential equations of the form $\dot{x} = X(x)$ are qualitatively equivalent if they have the same number of fixed points, of the same nature, arranged in the same order along the phase line. Note that the qualitative behaviour of x in the neighborhood of any isolated fixed point *must* correspond to one of phase portrait in figure 3.3, and this determines the *nature* of the fixed point. Thus, it is possible to divide the set of all differential equations of the form in equation 3.1 into equivalence classes, based on the relation of qualitative equivalence.

For non-linear X, there can be infinitely many different phase portraits. (the number of phase portraits depends on the number of fixed points and their locations). However, for linear X, we get $\dot{x} = ax$. Three cases are possible, depending on the values of a. If $a > 0$, $x = 0$ is a fixed point, and this point behaves as in figure 3.3(c). Similarly, the case $a < 0$ has a phase portrait as in figure 3.3(b). When $a = 0$, then every point x on the real line is a fixed point. Thus, the set of linear differential equations of the form $\dot{x} = ax$ can be classified into three distinct equivalence classes based on qualitative behaviour.

3.3.1.2   Two dimensional case

Consider the differential equation

$$\dot{\mathbf{x}}(t) = \frac{d\mathbf{x}}{dt} = \mathbf{X}(\mathbf{x}) \tag{3.3}$$

where $\mathbf{x} = (x_1, x_2)$ is a vector in R^2. This equation is equivalent to the system of two coupled equations

$$\begin{aligned} \dot{x}_1 &= X_1(x_1, x_2) \\ \dot{x}_2 &= X_2(x_1, x_2), \end{aligned} \tag{3.4}$$

where $\mathbf{X}(\mathbf{x}) = (X_1(x_1, x_2), X_2(x_1, x_2))$, because $\dot{\mathbf{x}} = (\dot{x}_1, \dot{x}_2)$.

Let us consider equation 3.4 from a geometric point of view. We consider two functions $x_1(t)$, $x_2(t)$ as specifying an unknown curve $x(t) = (x_1(t), x_2(t))$ in the (x_1, x_2) plane $\mathbf{R}^2$. Thus, x is a map from the real numbers $\mathbf{R}$ into $\mathbf{R}^2$. The right hand side of equation 3.4 expresses the tangent vector to the curve. The qualitative behaviour of the solutions is determined by how $x_1(t)$ and $x_2(t)$ behave as t increases. This geometrical representation of the qualitative behaviour of $\dot{\mathbf{x}} = \mathbf{X}(\mathbf{x})$ is called its *phase portrait*. The phase portrait records only the *direction* of the velocity of the phase point, and therefore represents the dynamics in a qualitative way. The phase portrait is therefore a two dimensional figure and the qualitative behaviour is represented by a family of curves, directed with increasing t, known as *trajectories*.

To examine the qualitative behaviour in the plane, we must look at the *fixed points* of equation 3.3. A solution $\mathbf{x}(t) = \mathbf{c} = (c_1, c_2)$ of equation 3.3 is called a *fixed point* if $\mathbf{X}(\mathbf{c}) = \mathbf{0}$. These solutions are called fixed points of the equation because $\mathbf{x}$ remains at $\mathbf{c}$ for all t.

Let us consider an example of an isolated fixed point, and look at a particular set of trajectories. Consider the system

$$\dot{x}_1 = -x_1, \quad \dot{x}_2 = -x_2 \tag{3.5}$$

This has a fixed point at (0,0), and solutions

$$x_1(t) = C_1 e^{-t}, \quad x_2(t) = C_2 e^{-t}, \tag{3.6}$$

where C_1 and C_2 are real constants. From equation 3.6,

$$x_1(t) = K x_2(t) \tag{3.7}$$

where $K = C_2/C_1$. Figure 3.4 shows the qualitative behaviour of this system. The family of trajectories is a set of radial straight lines passing

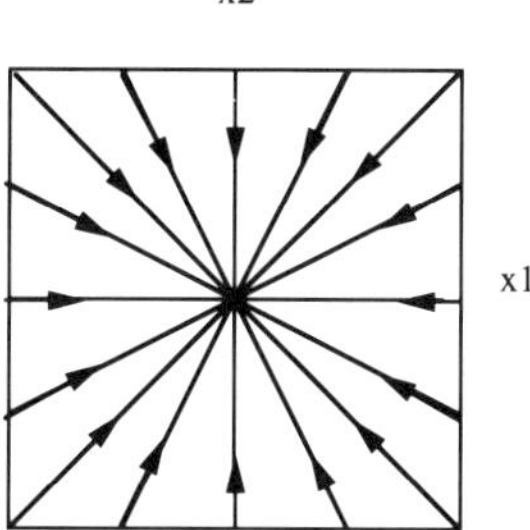

FIGURE 3.4. Illustrating the phase portrait for the system of equations in equation 3.5

through the origin. Furthermore, for any choice of C_1 and C_2, $\mid x_1(t) \mid$ and $\mid x_2(t) \mid \rightarrow 0$ as $t \rightarrow \infty$. This is indicated by the direction of the arrow on the trajectory, ie if $(x_1(t), x_2(t))$ is the starting point, then this *phase point* will move along the radial straight line towards the origin as t increases.

Figures 3.5(a)-(d) show phase portraits for other systems of differential equations. Note that qualitatively different solutions $(x_1(t), x_2(t))$ lead to trajectories with different geometrical properties. At this point it is important to understand what is meant by the term *qualitative difference*. In figures 3.5(a) and (b), all the trajectories are directed towards the origin, Hence this is the dominant qualitative feature, and the exact shape of the trajectories is unimportant. By the same token, figures 3.5(a) and (c) are qualitatively different. This intuitive notion will be made clearer when we examine linear systems in greater detail.

In fact, there are infinitely many qualitatively different planar phase portraits containing a single fixed point. However, for a linear system of equations, there is only a finite number of qualitatively different phase portraits. This will be investigated in the next section.

3.3.2   LINEAR SYSTEMS

A system

$$\dot{\mathbf{x}}(t) = \frac{d\mathbf{x}}{dt} = \mathbf{X}(\mathbf{x}) \tag{3.8}$$

where $\mathbf{x}$ is a vector in R^n is called a *linear system* of dimension n if $X : R^n \rightarrow R^n$ is a linear mapping. It can be shown that only a finite number of qualitatively different phase portraits can arise for linear systems [7]. If $X : R^n \rightarrow R^n$ is a linear mapping, equation 3.8 can be written in the form

$$\dot{\mathbf{x}}(t) = \mathbf{X}(\mathbf{x}) = \mathbf{A}\mathbf{x} \tag{3.9}$$

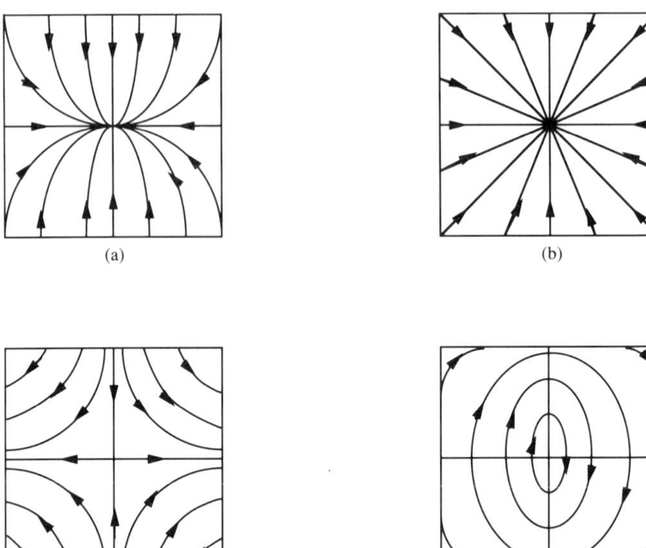

FIGURE 3.5. Illustrating the phase portraits for different systems of equations

where $\mathbf{A}$ is the coefficient matrix. Consider the effect of making a linear change of variables

$$\mathbf{x} = \mathbf{M}\mathbf{y} \tag{3.10}$$

where $\mathbf{M}$ is non-singular. Thus

$$\dot{\mathbf{y}} = \mathbf{M}^{-1}\mathbf{A}\mathbf{M}\mathbf{y} = \mathbf{B}\mathbf{y} \tag{3.11}$$

with

$$\mathbf{B} = \mathbf{M}^{-1}\mathbf{A}\mathbf{M} \tag{3.12}$$

Thus the coefficient matrix $\mathbf{B}$ of the transformed system is similar to $\mathbf{A}$. Since similarity is an equivalence relation on the set of $n \times n$ real matrices, this set can be divided into equivalence classes under the relation of similarity. Thus, if the solution to one system in a given equivalence class can be found, the solution for other members of the equivalence class can be generated through a similarity transformation.

Consider the set of $n \times n$ real matrices. The similarity classes for this set can be grouped into just finitely many types, known as the Jordan canonical forms for the matrices. The Jordan forms $\mathbf{J}$ for a real 2×2 matrix are as follows:

$$(a) \begin{bmatrix} \lambda_1 & 0 \\ 0 & \lambda_2 \end{bmatrix} \qquad (b) \begin{bmatrix} \lambda_0 & 0 \\ 0 & \lambda_0 \end{bmatrix}$$

$$(c) \begin{bmatrix} \lambda_0 & 1 \\ 0 & \lambda_0 \end{bmatrix} \qquad (d) \begin{bmatrix} \alpha & -\beta \\ \beta & \alpha \end{bmatrix} \tag{3.13}$$

where $\lambda_1 > \lambda_2, \beta > 0$, and λ_0, λ_1, λ_2, α, and β are real numbers.

When $\mathbf{J} = \mathbf{M}^{-1}\mathbf{A}\mathbf{M}$, $\mathbf{M}$ non-singular, then $\mathbf{J}$ is said to be the Jordan form of $\mathbf{A}$. The eigenvalues of the matrix $\mathbf{A}$ are

$$\lambda_1 = \frac{1}{2}(tr(\mathbf{A}) + \sqrt{\Delta}) \;\; and \;\; \lambda_2 = \frac{1}{2}(tr(\mathbf{A}) - \sqrt{\Delta}) \tag{3.14}$$

where $tr(\mathbf{A})$ is the trace, $det(\mathbf{A})$ is the determinant, and

$$\Delta = (tr(\mathbf{A}))^2 - 4\,det(\mathbf{A}) \tag{3.15}$$

The nature of the eigenvalues determines the Jordan form, and ultimately the qualitative behaviour of the phase portrait of the underlying linear differential equation. We shall now present phase portraits for the different Jordan forms described above.

There are two broad categories, depending on the nature of the matrix $\mathbf{A}$.

3.3.2.1   Case 1: Matrix $\mathbf{A}$ is non-singular

In this case $\mathbf{A}$ has non-zero eigenvalues, and the only solution to $\mathbf{A}\mathbf{x} = \mathbf{0}$ is $\mathbf{x} = \mathbf{0}$. Thus, for non-singular $\mathbf{A}$, there is only one isolated fixed point at the origin.

Figure 3.6 summarizes the different phase portraits that are possible. The phase portraits are shown in a neighborhood of the origin. The nomenclature for the types of phase portrait, viz. *node, saddle, star-node, improper node, center* and *spiral* are standard terms in the geometric theory of differential equations [76], [7].

When all trajectories point towards the origin the fixed point is said to be *stable*, and the converse holds for an *unstable* fixed point. However, for the purposes of this chapter we shall not be concerned with the notion of stability.

3.3.2.2   Case 2: Matrix $\mathbf{A}$ is singular

In this case, there are non-trivial solutions to $\mathbf{A}\mathbf{x} = \mathbf{0}$ and the system has fixed points other than the origin. For a two dimensional linear system, if $\mathbf{A}$ is singular, it has either a rank of 1 or is null. In the first case, there is a line of fixed points passing through the origin, and in the second case, every point in the plane is a fixed point. Figure 3.7 illustrates one particular case when the matrix $\mathbf{A}$ is singular.

Figure 3.8 depicts a graphical classification of phase portraits in the 2×2 linear case based on the nature of the eigenvalues. The interpretation of this figure is as follows. Since the eigenvalues depend on the trace and determinant, one can deduce the nature of a phase portrait from its associated trace and determinant. The curve $\Delta = 0$ is an important partitioning curve in the trace-determinant space due to equation 3.15.

CASE	JORDAN FORM	TYPE OF PHASE PORTRAIT	APPEARANCE OF PHASE PORTRAIT
1) Real distinct eigenvalues $\lambda 1,\ \lambda 2$ with $\lambda 1 > \lambda 2$	(a) $\begin{bmatrix} \lambda 1 & 0 \\ 0 & \lambda 2 \end{bmatrix}$ $\lambda 1$ and $\lambda 2$ have the same sign	NODE	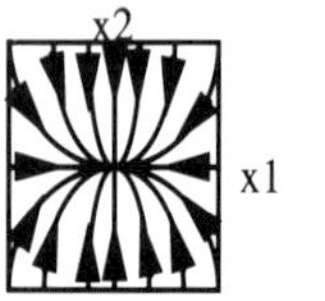
	(b) $\begin{bmatrix} \lambda 1 & 0 \\ 0 & \lambda 2 \end{bmatrix}$ $\lambda 1$ and $\lambda 2$ have opposite sign	SADDLE	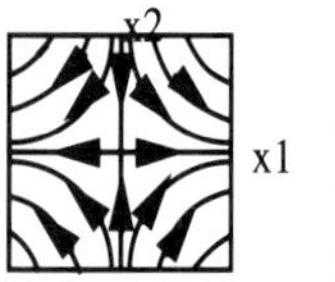
2) Equal eigenvalues $\lambda 1\ =\lambda 2 = \lambda 0$	(a) $\begin{bmatrix} \lambda 0 & 0 \\ 0 & \lambda 0 \end{bmatrix}$	STAR-NODE	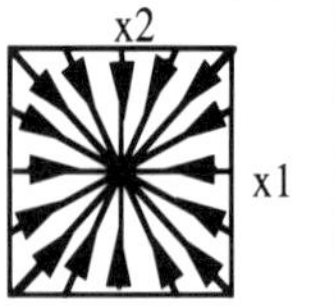
	(b) $\begin{bmatrix} \lambda 0 & 1 \\ 0 & \lambda 0 \end{bmatrix}$	IMPROPER NODE	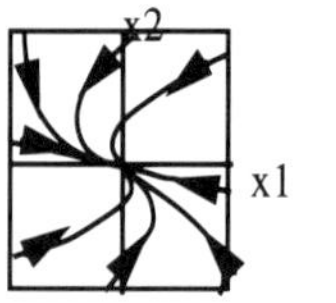
3) Complex eigenvalues $\lambda 1 = \alpha + \iota\beta$ $\lambda 2 = \alpha - \iota\beta$ $\alpha = \dfrac{1}{2}\ \mathrm{tr}(A)$ $\beta = \dfrac{1}{2}\sqrt{-\Delta}$	$\begin{bmatrix} \alpha & -\beta \\ \beta & \alpha \end{bmatrix}$ (a) $\alpha = 0$	CENTER	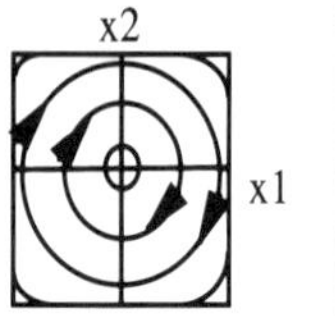
	(b) $\alpha \neq 0$	SPIRAL	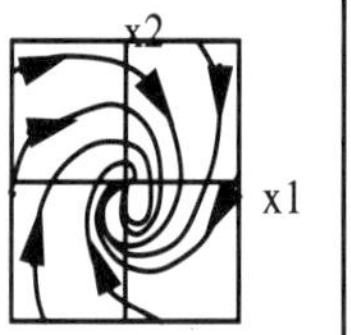

FIGURE 3.6. A classification of different phase portraits based on the nature of eigenvalues and the associated Jordan forms

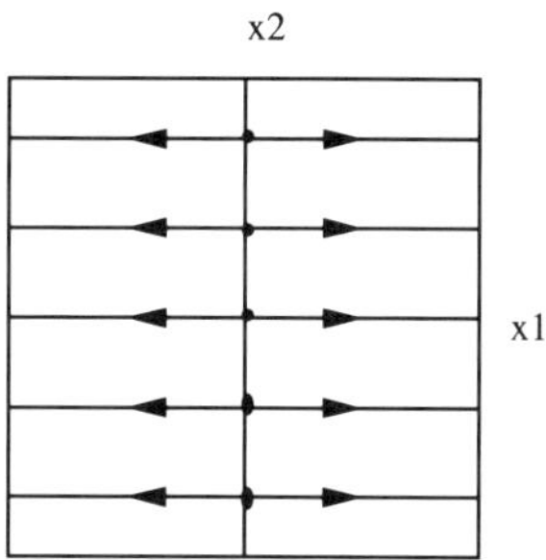

FIGURE 3.7. A singular phase portrait

The phase portrait when $\mathbf{A}$ is singular. In this case,

$$\mathbf{A} = \left[\begin{array}{cc} 1 & 0 \\ 0 & 0 \end{array} \right].$$

Every point on the x_2 axis is a fixed point.

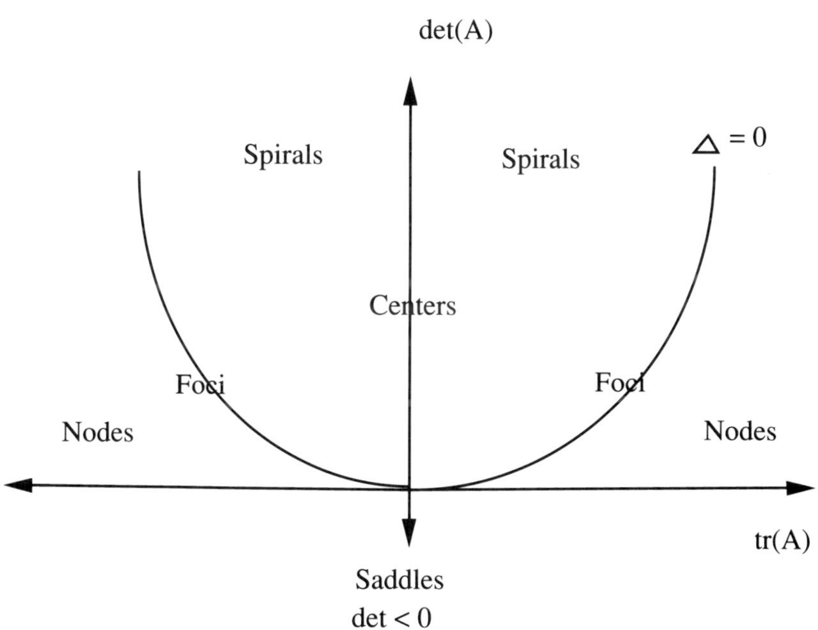

FIGURE 3.8. Summary of phase portraits for the two dimensional case

3.3.2.3 Affine transformations

Let us consider the case of an affine transformation instead of a linear transformation. In this case, we get

$$\dot{\mathbf{x}}(t) = \mathbf{X}(\mathbf{x}) = \mathbf{A}\mathbf{x} + \mathbf{b} \tag{3.16}$$

where $\mathbf{b}$ is a constant vector. In order to analyze this situation let us assume that $\mathbf{A}$ is nonsingular. Then the above equation has a fixed point at

$$\mathbf{x_0} = -\mathbf{A}^{-1}\mathbf{b} \tag{3.17}$$

The substitution of $\mathbf{y} = \mathbf{x} - \mathbf{A}^{-1}\mathbf{b}$ into equation 3.16 gives

$$\dot{\mathbf{y}} = \mathbf{A}\mathbf{y} \tag{3.18}$$

which has a fixed point at the origin. Since equation 3.18 is of a familiar form that we have studied, the discussion of phase portraits in the previous section still holds. The advantage in using affine transformations is that an isolated fixed point can lie anywhere in the phase plane, and need not be constrained to be at the origin. This will become clearer in section 3.4.

3.3.3 RELEVANCE OF THE THEORY

To see how the theory we have just discussed is relevant, we must make a connection between an actual texture and the qualitative or geometric theory of differential equations. This connection is afforded by the concept of *flow* [51]. We have looked at the phase portraits of different differential equations so far. These portraits represent the solution curves of the equations. One can think of points of R^n as flowing simultaneously along these solution curves, and this is called the *flow* of the differential equation. The phase portraits give a good visualization of the corresponding flows. The flow is linear because we are looking at linear differential equations.

An arbitrary oriented texture could have a highly complex global phase portrait, with several fixed points. A reasonable approach towards analyzing such textures is to restrict the global phase portrait to neighborhoods, to which a linear approximation can be applied. This is called *linearization* and is discussed in [7]. Thus, a complex oriented texture can be broken down into separate regions, such that in each region, the equation $\dot{\mathbf{x}}(t) = \mathbf{X}(\mathbf{x})$ can be linearized.

In this manner, one can treat a texture as being comprised of *piecewise* linear flows. What we must accomplish is to uncover the differential equation, or the phase portrait that best matches a portion of the given texture. This will allow for a qualitative description of the texture which is firmly grounded in the theory of differential equations. The details of the process are explained in the section 3.4.

At this point, we wish to clarify the notion of "qualitative". Andronov *et al* [5] provide a mathematical definition of "qualitative" properties in the context of dynamic systems to be those properties of paths, sets of paths and phase portraits which are preserved under arbitrary topological mappings of the region or set. (A topological mapping of a plane onto itself can be intuitively described as a rubber-sheet deformation of the plane by stretching or squeezing but without tearing or folding). From this perspective, one is motivated to describe an oriented texture such that the description remains invariant under a topological mapping. One such description consists of assigning a linear phase portrait to the given texture, as discussed in section 3.4. For instance, if a given texture is described as being a 'spiral phase portrait', it does not matter what the orientation and winding number of the spiral are. Thus, the quality assigned to a local region of a texture is the type of linear phase portrait that best approximates the region.

The reason we make the assumption of linearity is because an arbitrary oriented texture could have a highly complex global phase portrait. A reasonable approach towards analyzing such textures is to restrict the global phase portrait to neighborhoods, to which a linear approximation can be applied. This is called *linearization*, and is discussed in [7]. The linearization approach works in practice due to the Grobman-Hartman theorem [95]. This states that in the neighborhood of a hyperbolic critical point (one whose linearization has eigenvalues with non-zero real parts) the linearized flow field is homeomorphic to (i.e. qualitatively the same as) the original flow field in a neighborhood of the critical point.

3.3.4 THE PERCEPTUAL SIGNIFICANCE OF FLOW-LIKE PATTERNS

The initial experiment by Glass [38] showed that circular patterns are immediately perceived when a random dot pattern is superimposed on itself and rotated slightly. There are no apparent contours in the superimposed pattern – only a strong impression of flow. This led to the speculation that the visual system may perform local autocorrelations and integrate these to form global percepts (a circle for instance).

A later experiment by Glass and Perez [39] demonstrated that random dot interference patterns gain perceptual significance when they correspond to the trajectories of a system of linear differential equations. The specific examples in the experiment used correspond to center, star-node, spiral and saddle phase portraits. This strongly suggests that phase portraits constitute perceptually meaningful descriptions of flow.

Stevens [111] showed that local orientation structure can be discerned for composite dot patterns that are created by combining portions of the Glass and Perez patterns [39]. Specifically, he illustrates the case of a portion of a spiral phase portrait adjoining a portion of a node phase portrait.

This supports our approach of fitting piecewise linear flows to the orientation field. Stevens [111] also developed an algorithm for recovering local orientation organization in Glass patterns. The output of this algorithm consists of line segments connecting pairs of dots, and is analogous to the orientation fields that we consider in this paper.

An alternate approach to extracting perceptually meaningful structure from dot patterns is to use graph theoretic methods. Zahn [133] incorporates Gestalt principles into a minimal spanning tree algorithm in order to detect and describe clusters in dot patterns. Toussaint [117] suggests the use of a relative neighborhood graph as a basis for perceptual organization. Yip [130] uses minimal spanning trees to group discrete points into phase portraits.

Zucker [136] models the orientation selection problem as one of deriving a vector field of tangents. He addresses perceptual issues related to the derivation of this tangent field, such as the effect of size and spacing of dots that form contours. However, the problem of understanding the geometric structure of the tangent field was not dealt with.

Interestingly, though all the research above shows that there is some sort of perceptual grouping of the entities of a vector field, there is no precise characterization for *how* the grouping should proceed. Hence the research in this paper attempts to fill this void by suggesting that one mechanism for grouping is to use linear phase portraits. There are two caveats at this point. First, there is nothing special about using *linear* phase portraits. The linear approximation is being made to render the problem tractable. Second, no claims are being made about the existence of phase portrait grouping mechanisms in the human visual system. Our approach is driven from a purely computational viewpoint. Further experiments on the human perception of phase portraits need to be carried out before such a claim can be established.

3.4   Experimental Methods

In this section we examine the details of how one can apply the qualitative theory of differential equations to real images. The basic scenario is illustrated in figure 3.9. We have a given oriented pattern, in the form of oriented line segments. Such an output can be obtained from the orientation estimation algorithm described in section 3.2. The question we address is: which phase portrait for a 2×2 linear system most closely resembles that in figure 3.9(a)? The resulting answer is displayed in figure 3.9(b). This is the basic idea behind the qualitative description of textures.

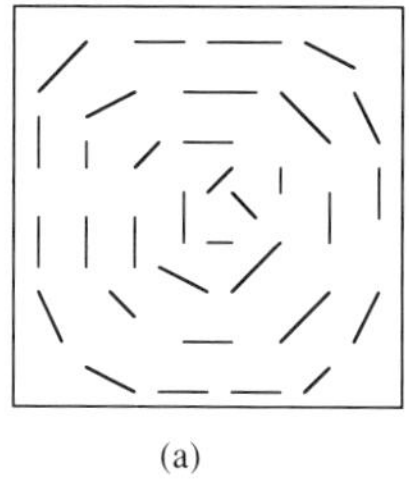
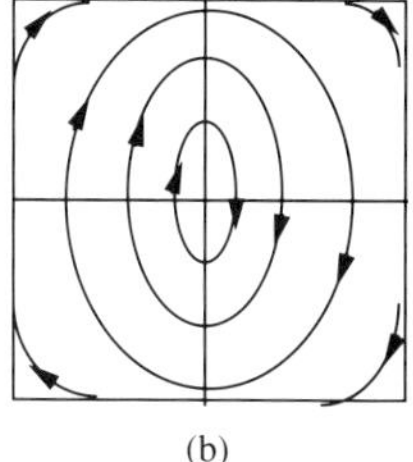

(a)   (b)

FIGURE 3.9. (a) A given orientation field. (b) The linear phase portrait which resembles the orientation field.

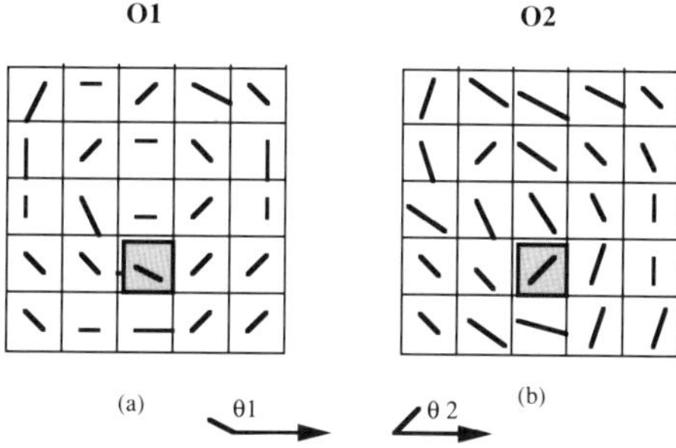

FIGURE 3.10. Two orientation fields

3.4.1 A DISTANCE MEASURE

We need some measure of the distance between two orientation fields. Since we are dealing with orientation fields, we must remember the caveat in section 3.2.4, which identifies the inherent ambiguities present in such a situation.

Let us consider two discrete orientation fields O_1 and O_2, as in figure 3.10. At each point (i, j), the orientation field is represented by a line segment, which has length equal to the coherence at that point, and an ori-

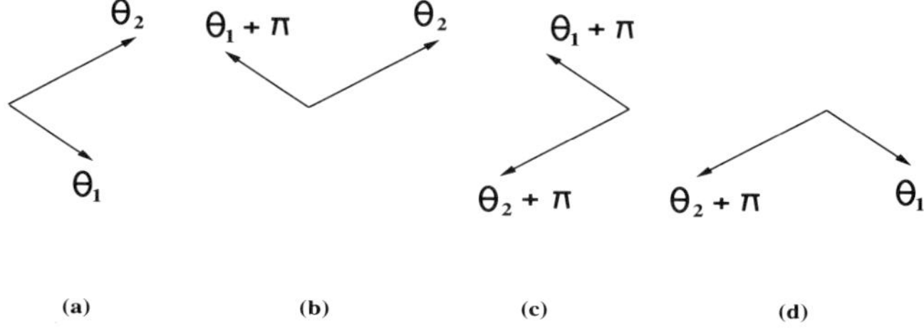

(a)   (b)   (c)   (d)

FIGURE 3.11. Possible cases for ambiguity

entation which is the dominant local orientation of the underlying texture. Let $R_{k_{ij}}$ represent the length of a segment in the orientation field $\mathbf{O_k}$ (k is either 0 or 1) at location (i, j), and let $\theta_{k_{ij}}$ be the angle subtended by this segment. Let us examine the segments at identical locations in the two fields, as shown in figure 3.11.

Since the orientation of each of the segments is ambiguous in sign, we could have four possible cases, as shown. The distance measure we choose should be *independent* of these sign ambiguities. Hence, for instance it would be incorrect to take the dot product of the two oriented segments. Instead, the measure that we use is the area of the triangle formed by the oriented segments, which is

$$A_{ij} = \frac{1}{2} R_{1_{ij}} R_{2_{ij}} \mid sin(\theta_{1_{ij}} - \theta_{2_{ij}}) \mid \tag{3.19}$$

The area is a measure of the difference or disparity between two oriented segments; the area becomes zero when the two segments have the same orientation. In fact, the disparity varies as $sin\,\theta$, which is interesting because even though the two vectors may point in opposite directions, the measure is zero.

We sum these differences over the entire field to obtain

$$S = \sum_{i,j \in W} A_{ij} \tag{3.20}$$

where W is the region over which the fields exist. The smaller S is, the closer the two fields are.

3.4.2 Non-linear least squares fitting

The problem we wish to tackle is to find a linear 2D phase portrait that 'best' describes a given orientation field. We define the best estimate to be *one that minimizes the sum S, as defined in equations 3.19 and 3.20.* The problem viewed in this fashion becomes a non-linear least squares problem.

It should be mentioned that this approach is not the same as matching one of the canonical phase portraits to the given texture. Instead, the solution to the optimization problem is found over a continuous parameter space, and the result is used to classify the underlying flow into a fixed number of classes. The reason for this is that each class of phase portraits has an infinite number of members present in it, all of which are qualitatively equivalent.

We employ the Levenberg-Marquardt technique to solve the above non-linear least squares problem. Specifically, we used the function LMDIF, a Levenberg-Marquardt non-linear least squares solver from MINPACK, a minimization algorithms package [37, 91]. The parameters are set up for MINPACK as follows. We begin with an initial guess for the matrices $\mathbf{A}$

and $\mathbf{b}$. This is the predicted model for the phase portrait. The predicted model gives rise to a predicted orientation field $\mathbf{O_2}$,[3] which is compared with the actual orientation field $\mathbf{O_1}$, according to equations 3.19 and 3.20. The Levenberg-Marquardt method converges to a solution $\hat{\mathbf{A}}$ and $\hat{\mathbf{b}}$, which are best estimates in a least squares sense.

3.4.3 IMPLEMENTATION DETAILS

We discuss some issues in the implementation of the non-linear least squares technique.

3.4.3.1 Generating the orientation field $\mathbf{O_2}$

The orientation field $\mathbf{O_2}$ is generated by taking an odd sized square window of size NxN pixels, locating the origin of the coordinate system at the center of the window, and generating flow vectors at each point in the window. The row and column values correspond to the variables x_1 and x_2 in equation 3.4, and are plugged into equation 3.16 to generate the values of $\dot{x}_1$ and $\dot{x}_2$. The vector $\dot{\mathbf{x}}$ is then normalized, to give a unit vector pointing in the direction of flow. This operation is performed at each pixel in the window.

3.4.3.2 Normalization of orientation fields

We normalize both the orientation fields $\mathbf{O_1}$ and $\mathbf{O_2}$ (i.e. we consider only unit vectors). There are two reasons for doing this.

In many physical situations, the coherence of an image may not have a direct physical correlate, since it is primarily a visual texture. For instance, the coherence need not be proportional to speed, in the case of fluid flow. This implies that we must omit the magnitude of the vectors from equation 3.19 and use only the *direction* information.

Secondly, the model $\dot{\mathbf{x}}(t) = \mathbf{X}(\mathbf{x}) = \mathbf{A}\mathbf{x} + \mathbf{b}$ may not specify the magnitudes in the vector field as we desire. However, it *does* specify the directions of the vectors correctly.

Hence, we set $R_{1_{ij}} = R_{2_{ij}} = 1$ in equation 3.19. Another approach would be to retain the coherence of the original image, and normalize the predicted orientation field. This is useful if one wants to perform quantitative measurements on the flow field, where the coherence has a direct physical interpretation (say the coherence is proportional to the concentration of fluid at a point in space). The issue of retaining coherence information is domain dependent. Nevertheless, only trivial modifications are needed in equation 3.19, which defines the term to be minimized. Since our primary

---

[3]Actually it gives rise to a vector field, but we treat it as an orientation field because there is still a *global* sign ambiguity

goal at this stage is to obtain symbolic descriptors for the flow we normalize both orientation fields.

3.4.3.3   Uniqueness of the solution

Scaling $\mathbf{A}$ and $\mathbf{b}$ by an arbitrary factor changes only the magnitude of the vector at a point, but not its direction. Since we utilize only the directional information, this implies that the solution to the least-squares problem is not unique. If $\hat{\mathbf{A}}$ and $\hat{b}$ are optimal solutions to the non-linear least squares problem, then $k\hat{\mathbf{A}}$ and $k\hat{b}$ are also optimal solutions, where k is an arbitrary scale factor. Hence, we would like to make the solution unique, so that we can obtain a numerical solution to the least squares problem.

In our implementation, uniqueness is enforced as follows. The matrix $\mathbf{A}$ can be regarded as a 4 dimensional vector of the form (x_1, x_2, x_3, x_4). Denote this 4-dimensional vector space by $\mathbf{W}$. Let $\mathbf{W}_1$ and $\mathbf{W}_2$ be two affine subspaces of $\mathbf{W}$ such that $\mathbf{W}_1$ consists of vectors of the form $(1, x_2, x_3, x_4)$, and $\mathbf{W}_2$ consists of vectors of the form $(0, 1, x_3, x_4)$.

In other words, we are forcing the leading term in the matrix $\mathbf{A}$ to be either 1 or 0. This automatically takes care of the scaling problem, because we look for solutions whose first component is either 1 or 0. Solutions whose first component is different from 1 or 0 can be generated by a simple scaling. Note that in the subspace $\mathbf{W}_2$, the first non-zero component is normalized to be 1.

We search for solutions in the two affine subspaces $\mathbf{W}_1$ and $\mathbf{W}_2$, and choose the best of these two solutions. To find a solution in $\mathbf{W}_1$, we used the starting values $\mathbf{A} = (1, 1, 0, 0)$ and $\mathbf{b} = 0$. For a solution in $\mathbf{W}_2$, we used the starting values $\mathbf{A} = (0, 1, 1, 0)$ and $\mathbf{b} = 0$. The method is not sensitive to the starting point, as our experiments demonstrated convergence from other starting points, such as $\mathbf{A} = (1, 0, 0, 1)$ (the identity matrix). These starting points were determined by trial and error as follows. We hypothesized a starting point and ran the algorithm on the different canonical phase portraits generated from equation 3.13. The only admissible starting points are those which converge to the correct solution in all cases.

The Levenberg-Marquardt method [91] uses steepest descent search far away from the solution and the Newton method close to the solution. Newton's method has the advantage of quadratic convergence in the vicinity of the solution.

Note that there is an ambiguity in the signs of the matrices $\mathbf{A}$ and $\mathbf{b}$. This arises because sign reversals of $\mathbf{A}$ and $\mathbf{b}$ simultaneously will give rise to the same result. This means that at best, our solution can be unique only upto a sign reversal factor. As we show in the next section, this does not cause problems with the symbolic description scheme and the technique for segmentation.

3.4.3.4   Real Images

We use the following technique to analyze orientation fields derived from real images. We sweep a window of some predetermined size (11 pixels in our experiments) over the entire image. At each location of the window, we perform a least squares fit to obtain the matrices $\mathbf{A}$ and $\mathbf{b}$ that best describe the flow. Once these matrices are obtained at each point, we can perform operations such as segmenting the image and locating the positions of central patterns, as described in the next section.

3.4.4   SEGMENTATION

Segmentation of the flow image into qualitatively different classes can be performed once we obtain an estimate of the form

$$\dot{\mathbf{x}}(t) = \hat{\mathbf{A}}\mathbf{x} + \hat{\mathbf{b}} \tag{3.21}$$

where $\hat{\mathbf{A}}$ and $\hat{\mathbf{b}}$ are least squares estimates. This estimate is obtained for a fixed window-size, which is predetermined. Once a window-size is selected, least squares fitting is performed over successive overlapping windows until the entire image is covered.

From $\hat{\mathbf{A}}$, the quantities $det(\hat{\mathbf{A}})$ and $tr(\hat{\mathbf{A}})$ are found, which determine the nature of the eigenvalues of $\hat{\mathbf{A}}$. This places the given local orientation field in the $det - tr$ plane as in figure 3.8 and hence classifies the underlying oriented pattern.

3.4.4.1   A measure for closeness of fit

We can also obtain a measure of the closeness of fit between a pattern of the form $\dot{\mathbf{x}}(t) = \hat{\mathbf{A}}\mathbf{x} + \hat{\mathbf{b}}$, and the given orientation field. This is readily obtained from equations 3.19 and 3.20 as follows. Consider a window $\mathbf{W}$, over which the given orientation field is $\mathbf{O_1}$, and let $\mathbf{O_2}$ be the orientation field derived from the estimated model $\dot{\mathbf{x}}(t) = \hat{\mathbf{A}}\mathbf{x} + \hat{\mathbf{b}}$.

Since we are normalizing both $\mathbf{O_1}$ and $\mathbf{O_2}$, observe that when the difference between the orientation fields is small, the sum S in equation 3.20 is small. Thus, the better the fit, the smaller is S. Also note from equation 3.19 that for normalized orientation fields, $R_{1_{ij}} = R_{2_{ij}} = 1$, and hence the maximum value for A_{ij} at a point occurs when the sin term becomes unity. Thus, the maximum error at any point is $1/2$, and the worst case error for the entire window is $N^2/2$, where N is the size of the square window over which the fit is performed. This can be used to obtain a normalized estimate for the closeness of fit C_W, as given by

$$C_W = 1 - \frac{S}{N^2/2} \tag{3.22}$$

Minor modifications of this estimate can be used when the orientation fields are not normalized.

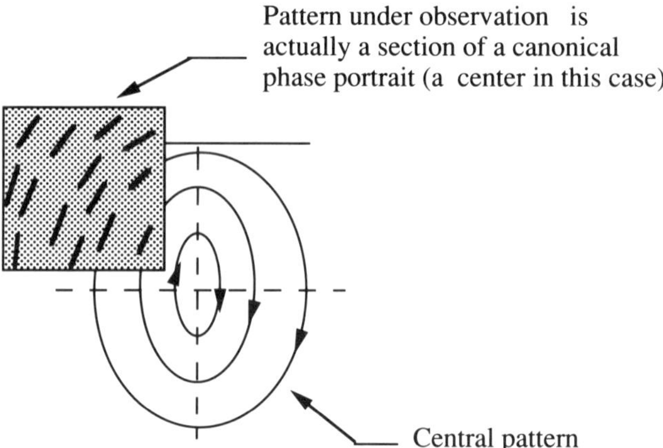

FIGURE 3.12. Illustrating the case when the pattern under observation is not a central pattern.

3.4.4.2 Selecting the size of the window

The basic criterion used in selecting the window-size is that we must choose a size that is sufficiently small to ensure linearity of the underlying flow. Of course, too small a window-size will result in poor estimates of the flow classes. Another factor that affects the choice of window-size is that larger windows are computationally more expensive. We do not have a method for automatically selecting the size of the window, and this is a topic we will be investigating in the future. One possibility is to select that window-size which gives rise to the smallest global error of fit, obtained by summing C in equation 3.22 over the entire image. Another possibility is to first segment the image using a given window-size, and to then use an iterative region growing algorithm to refine the segmentation, as done by Besl and Jain [14].

3.4.5 LOCATING FIXED POINTS

The reason for finding information about fixed points is that they constitute an important part of the qualitative description of flow-like textures. From a fluid dynamics point of view, fixed points of spiral patterns have the physical significance of being vortices. Such an interpretation is useful in analyzing turbulent flow [107].

Fixed points can be identified by noting that the term $\hat{b}$ need not be 0. This means that the portion of the oriented pattern over which we are performing the least square fit is actually a section of a larger linear flow pattern, whose fixed point is located away from the center of the current estimating window. This is illustrated in figure 3.12. The predicted fixed point of the orientation field is given by equation 3.17. This information can be used to estimate locations of *central patterns*, ie patterns that have the origin as a fixed point. A given texture may have several central patterns – eg saddles and spirals at different locations.

We perform the estimation of fixed points as follows. Suppose we find estimates $\hat{\mathbf{A}}$ and $\hat{\mathbf{b}}$ over a window $\mathbf{W}$, from which we determine the type of phase portrait, say P_i (i indexes the different types of phase portraits). We calculate the fixed point of the pattern as $\mathbf{x}_F = -\hat{\mathbf{A}}^{-1}\hat{\mathbf{b}}$. Using this, we update a measure V_{F,P_i} for the likelihood of a fixed point of type P_i being at $\mathbf{x}_F$ (the spirit of this approach is analogous to a Hough transform). For implementation purposes, we assume a discrete space for $\mathbf{x}_F$. Thus, if we observe different sections of the same spiral pattern, we keep updating a measure at the fixed point for that spiral pattern. This is done by incrementing the measure by the closeness of fit C_W, defined in equation 3.22 as follows

$$V_{F,P_i} \leftarrow V_{F,P_i} + C_W \tag{3.23}$$

The reason for this is that the 'vote' for a particular location should be proportional to the closeness of fit at the window which casts this 'vote'. This will suppress regions of poor fit from introducing incorrect hypotheses about the locations of the fixed points.

Thus, for each type of phase portrait P_i, we can generate a 2D map which shows the distribution of the likelihood of fixed points over the image plane. Thus, the value at a particular point on the map for P_i is proportional to the likelihood of that point being a fixed point of type P_i. For the sake of notation, we introduce the terms *saddle fixed point map*, *spiral fixed point map*, and *node fixed point map*.

3.4.6   RECONSTRUCTING THE ORIGINAL TEXTURE

We now describe the use of information about fixed points to reconstruct salient features in the original texture. This results in a symbolic description of the given texture.

The symbolic description is obtained from the saddle, node and spiral fixed point maps discussed in section 3.4.5. So far we have used only simple techniques to extract a symbolic description, but more robust techniques can easily be applied once the idea behind the method is understood. Our approach is to threshold the saddle, node and spiral fixed point maps at 50% of the maximum value in all three maps, and extract the centroids of the connected components. This is one scheme for automatically locating the fixed points of these types of phase portraits.

The justification for this approach is that fixed points that are visually dominant in the given texture will manifest themselves as regions of high likelihood in the fixed point maps. We seek to pinpoint the exact locations of these fixed points, and one way of doing this is to detect the centroids of clusters of high likelihood. Of course, there are a number of ways of doing this, such as two dimensional cluster analysis [58]. However the discussion of such methods falls outside the scope of this book. The use of such methods will improve the quality of our results, without necessitating a change in the philosophy of our approach.

The threshold is being used to filter out regions whose likelihood of being fixed points is low. The threshold has been arbitrarily chosen, and in fact can be varied with the following goal in mind. One might want to elucidate different descriptions of the same texture according to the salience of features present. The most dominant feature of the texture is obtained with a high threshold, and as the threshold is lowered, finer details begin to emerge. Thus, with a conservative (high) threshold, one may obtain only a few central patterns, and with a low threshold several central patterns may manifest themselves. This is analogous to the concept of scale or resolution, and a detailed discussion of this aspect is beyond the scope of this book.

Once the fixed points of various phase portraits are identified, we can find out the values of matrices $\hat{\mathbf{A}}$ and $\hat{b}$ that represent the best fit at each fixed point. This information can be used to reconstruct the given texture within a neighborhood of the fixed point. This is the method we use in order to reconstruct salient features of the original texture.

3.5   Experimental Results

We now present results of applying our algorithm to several real images. These images have been selected from the domains of fluid flow visualization, and lumber processing, mentioned in section 3.2.

For each test case, we obtain a segmentation of the given texture, and the computation of a likelihood map describing the locations of singularities or fixed points. These results are then used to derive symbolic descriptions of the given textures, and to reconstruct the original texture based on these descriptions.

3.5.1   FLOW PAST AN INCLINED PLATE

Figure 3.13a shows a 240x240 pixel image of flow past an inclined plate [118]. The result of applying the orientation estimation algorithm described in section 3.2 is displayed in figures 3.13b and 3.13c. The orientation field is calculated for each point in the image, but is displayed in a sampled form in order to avoid clutter.

Two methods of presenting the orientation field are used in order to provide easy means for interpretation. In the first method (figure 3.13b), the orientation field is overlayed on the original image to aid comparison. The orientation at each point is represented by means of a line segment, whose direction corresponds to the dominant local orientation, and whose length is proportional to the coherence. Observe that the segments capture the flow of the texture at each point, and orient themselves along the direction of flow.

In the second method (figure 3.13c), the orientation at each point in the image is encoded as the angle of orientation of a unit vector and is superimposed on the coherence which is encoded as a gray value. The brighter points indicate strong coherence of flow. The coherence image combined with the overlayed angle image provides a good description of the underlying flow-like texture.

Figure 3.13d shows the initially segmented flow image. Segmentation is performed into the following classes: nodes, spirals, and saddles (this is the smallest number of equivalence classes). The segmented regions are gray-level coded, according to the scheme shown in figure 3.13g. Figure 3.13e shows the segmented flow image after region refinement operations (described in appendix B) are performed on the initially segmented image. This is done in order to remove one-pixel sized holes and protrusions. This results in a cleaner segmented image. In all our subsequent results, we shall display only the refined segmented image.

Figure 3.13f shows the closeness of fit measure, C_W, over successive windows, as defined in equation 3.22. The measure C_W is quantized into gray levels, and displayed as an image, with the brighter intensity values corresponding to higher values of C_W.

Figures 3.13h and 3.13i show the saddle fixed point map and the spiral fixed point map, calculated from equation 3.23. The maps are printed such that the likelihood of a map point being a fixed point is proportional to the intensity at that point. These maps have been normalized with respect to each other, which means that normalization is performed with respect to the maximum likelihood over all types of fixed points.

The locations of these fixed points agrees with the nature of the original flow image. Furthermore, the fact that the saddle fixed points have higher intensity than the spiral fixed points means that the saddle-like nature of the original texture is more dominant than the spiral-like nature. (this could perhaps be verified through some experiments on human perception to compare what humans find dominant vis-a-vis what the algorithm finds dominant). We have not displayed the node fixed point map because the likelihood of node-fixed points was negligible compared to the likelihood of saddle and spiral fixed points (the image would have been virtually black if displayed).

The parameters used in segmentation are as follows. The original angle and coherence images were downsampled by a factor of 4 in order to reduce the computational burden. The non-linear least squares fitting was performed over a window of size 11x11. This window-size represented a balance between the factors of maintaining linearity, overcoming the effects of noise, and computation time, as mentioned in section 3.4.4.2. We have used the same set of parameters over all our experiments to prove that there is nothing mystical about the choice of this particular set. In fact, figure 3.14 shows the result of using a different windowsize, namely a window of 5x5 pixels in order to perform the least square fits. One can see that the qualitative nature of the segmented image in 3.14(a) is the same as in figure 3.13(d), though 3.14(a) appears to have more noise. Figures 3.14(b) and (c) show the saddle and spiral fixed point maps.

Reconstruction of flow past an inclined plate

The symbolic description of the image is automatically generated from the program as shown in figure 3.15(a). Figure 3.15(b) shows the reconstructed flow image based on values for $\mathbf{A}$ and $\mathbf{b}$ at the saddle fixed points.

The coordinates for the various fixed points are represented as row and column values, with the origin being at the upper left corner of the image. These images are of size 60x60, as the original image was downsampled by a factor of 4. The same convention holds for all the subsequent results presented.

3.5.2   Image of secondary streaming

Figure 3.16 analyzes secondary streaming induced by an oscillating cylinder [118, p. 23]. This pattern consists of four vortices around which there is circulation.

Reconstruction of secondary streaming

Figure 3.17 shows the symbolic description of the image in figure 3.16, and the reconstructed flow image based on values for $\mathbf{A}$ and $\mathbf{b}$ at the spiral fixed points.

An interesting aspect of the reconstructed image in figure 3.17 relates to the comment made in section 3.2.4. The problem in interpreting a raw orientation field is that the oriented segments do not have a unique direction, as we showed in figure 3.1. We have ameliorated this problem now because the phase portrait dictates the direction of flow at each point. Of course, there is still a *global* sign ambiguity, but we have now got rid of locally opposing flow vectors. Observe that in the reconstructed spirals, the flow around the spiral is coherent, and is either clockwise or anticlockwise. Compare figure 3.17 with figure 3.1 to observe this difference. This observation has important consequences if we want to derive a coherent *vector field* from a scalar field of observations.

3.5.3   Image of wood knots

Let us now consider an example from a different domain. The automation of lumber defect detection is vital for the future control of lumber processing, which is an important industry [113]. Defects in wood are rich in textural content, and typically involve the identification of knots, worm holes, checks, and grains [25, 96, 33, 23].

In order to illustrate the usefulness of our scheme, we have used it to

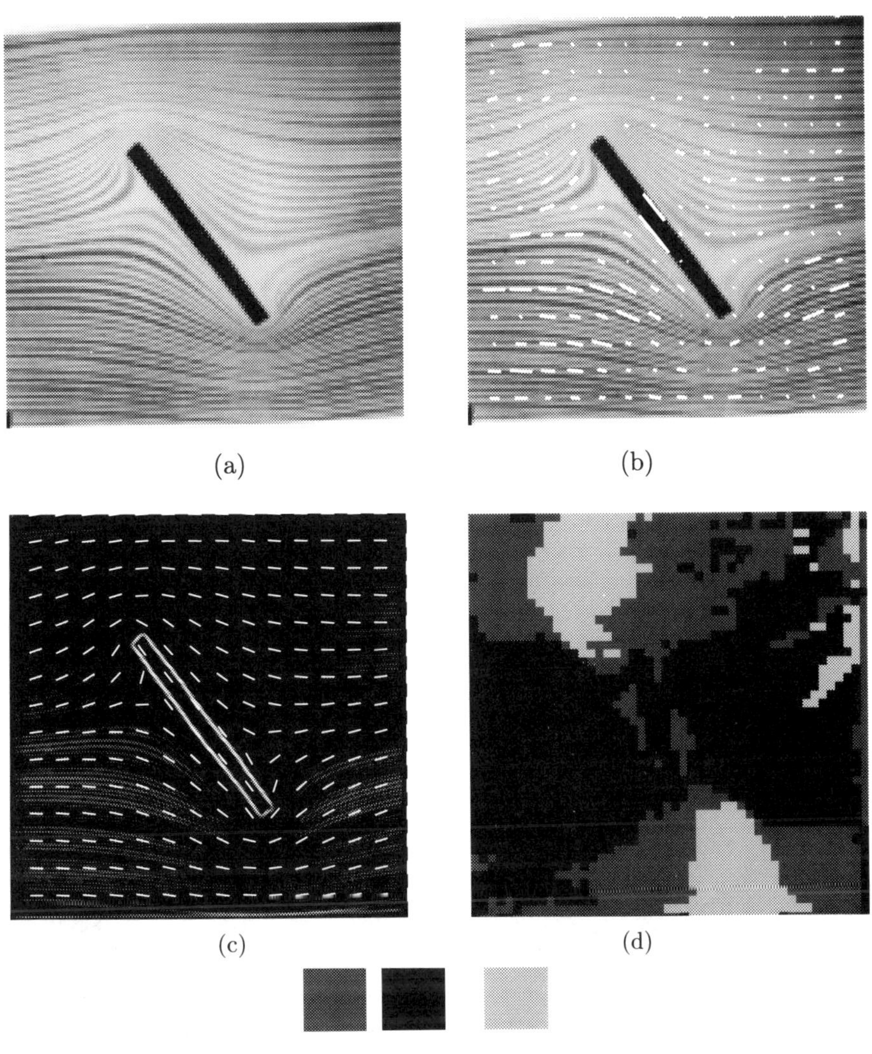

FIGURE 3.13. (a): The original flow image *(Photograph courtesy D. H. Peregrine)*. (b) The orientation field overlayed on the original image. The length of each oriented segment is proportional to the coherence. (c) The coherence image, with the orientation field overlayed. (d) The segmented image. Note that the areas to the right and left of the inclined plate are classified as saddles. This is what we should expect, because these areas indeed look like sections of saddles. Similarly, the areas around the top and bottom ends of the inclined plate are classified as spirals. Again, this makes sense because these areas seem like sections of spirals.

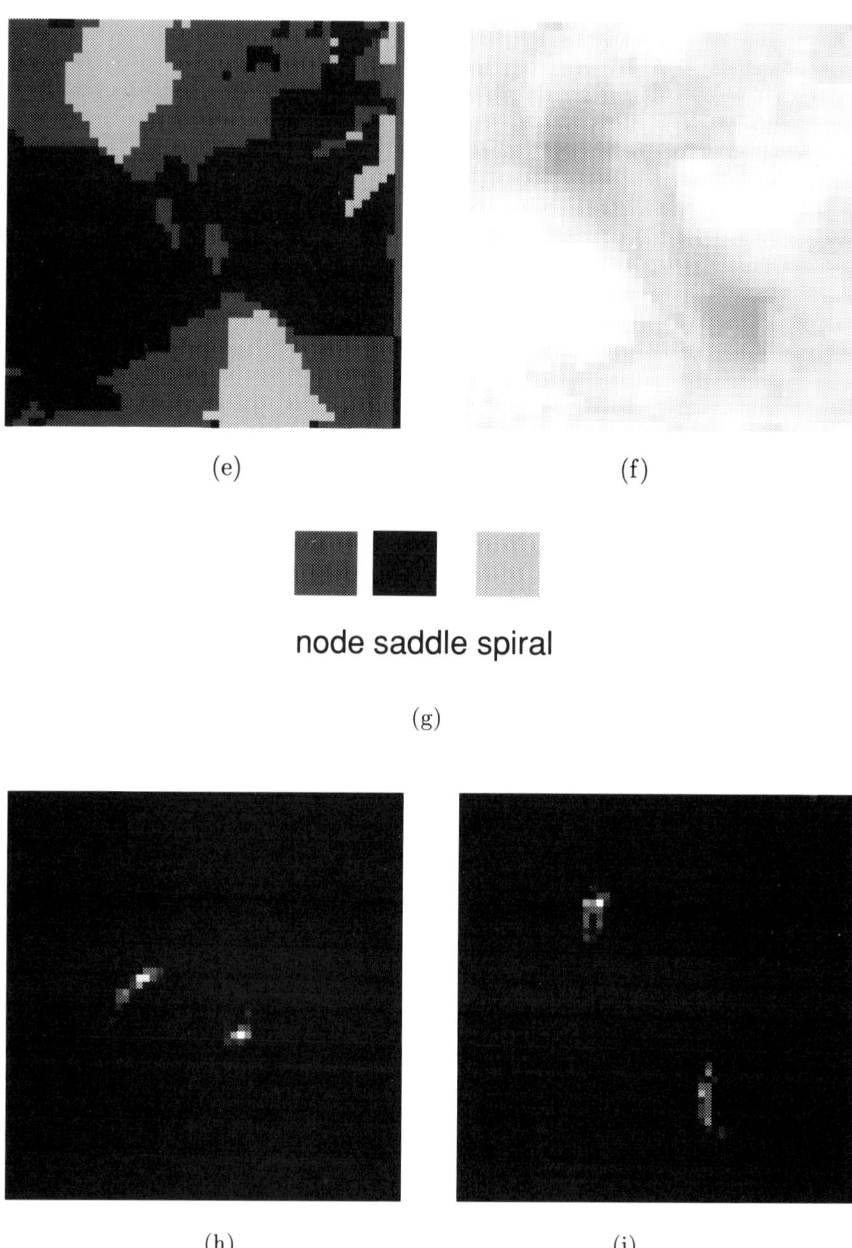

FIGURE 3.13. (e) The segmented image after region refinement. The different types of phase portraits possible have been coded. (f) The closeness of fit C_W, as defined in equation 3.22. (g) Shading key for phase portraits. (h) Map of saddle fixed points (i) Map of spiral fixed points

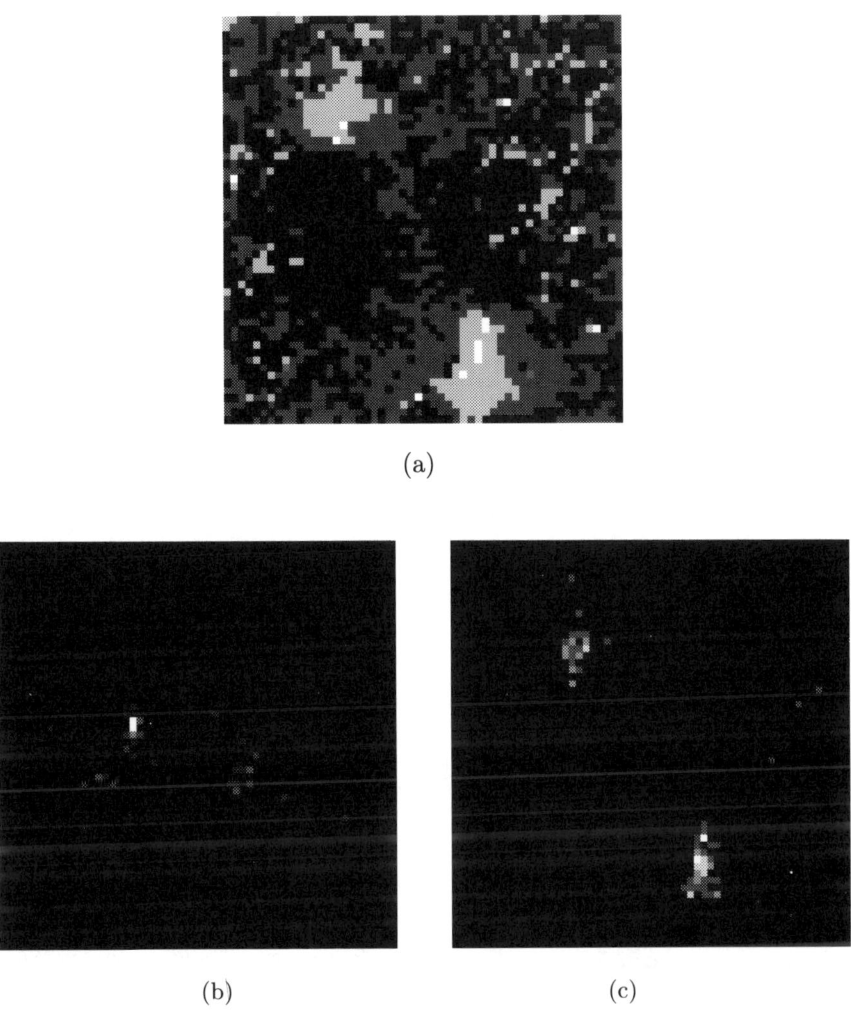

(a)

(b)   (c)

FIGURE 3.14. (a) The segmented image using a window size of 5x5 pixels to perform the least squares fit. (b) Map of saddle fixed points (c) Map of spiral fixed points

```
Description of given image:
 Saddle fixed points located
 at following points:-
(26, 19)     (34, 34)


 Values of A and B matrices are:


A = 1          0.362
     1.061    -0.802
B = -0.811    3.114 at (26, 19)


A =  1          0.217
     1.063    -0.777
B =  1.030    -2.012 at (34, 34)
```

FIGURE 3.15. (a): This is the symbolic description of the input image. (b): The reconstructed flow image, based on this symbolic description. This corresponds to the original image in figure 3.13 Observe that the salient feature of the original image, namely the saddle-like nature of the texture has been captured in the reconstructed image.

identify some of the defects that occur in wood such as knots. In fact, we have now have a rich representation of a knot, based on the notion of phase portraits. For instance, we can distinguish between different kinds of knots, eg. elliptical knots, circular knots and so on. This arises because knots fall under the category of spiral phase portraits, and the deformation and vorticity of the spiral phase portrait give us a means for distinguishing different kinds of spiral patterns.

Figure 3.18 analyzes two knots in a piece of wood, obtained from [33].

Reconstruction of wood knots

Figure 3.19 presents a symbolic description of the image in figure 3.18, and its reconstruction.

3.5.4   IMAGE OF COMPLEX FLOW

Figure 3.20a, obtained from [26] shows measurements of the concentration field $c(\mathbf{x}, t)$ of a conserved scalar in the self-similar far field of an axisymmetric turbulent jet. These data were obtained by imaging of the laser induced flourescence from a passive dye mixed with the jet fluid. The gray level at each point gives the instantaneous local concentration value.

Reconstruction

Figure 3.21 shows the symbolic description of the image in figure 3.20 and its reconstruction.

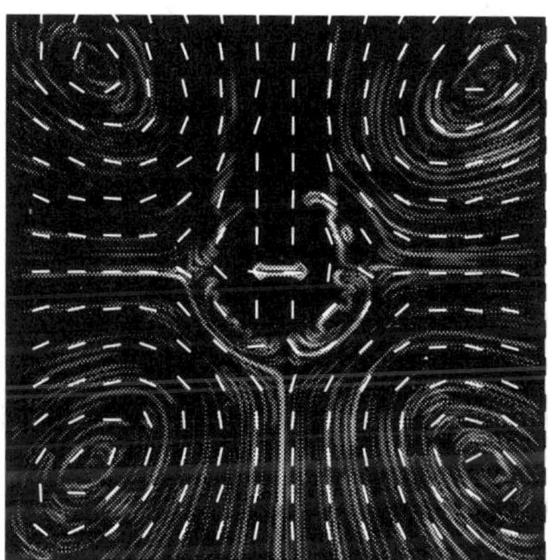

FIGURE 3.16. (a): The original flow image *(Photograph courtesy M. Van Dyke)*. (b) The orientation field overlayed on the original image. The length of each oriented segment is proportional to the coherence. (c) The coherence image, with the orientation field overlayed.

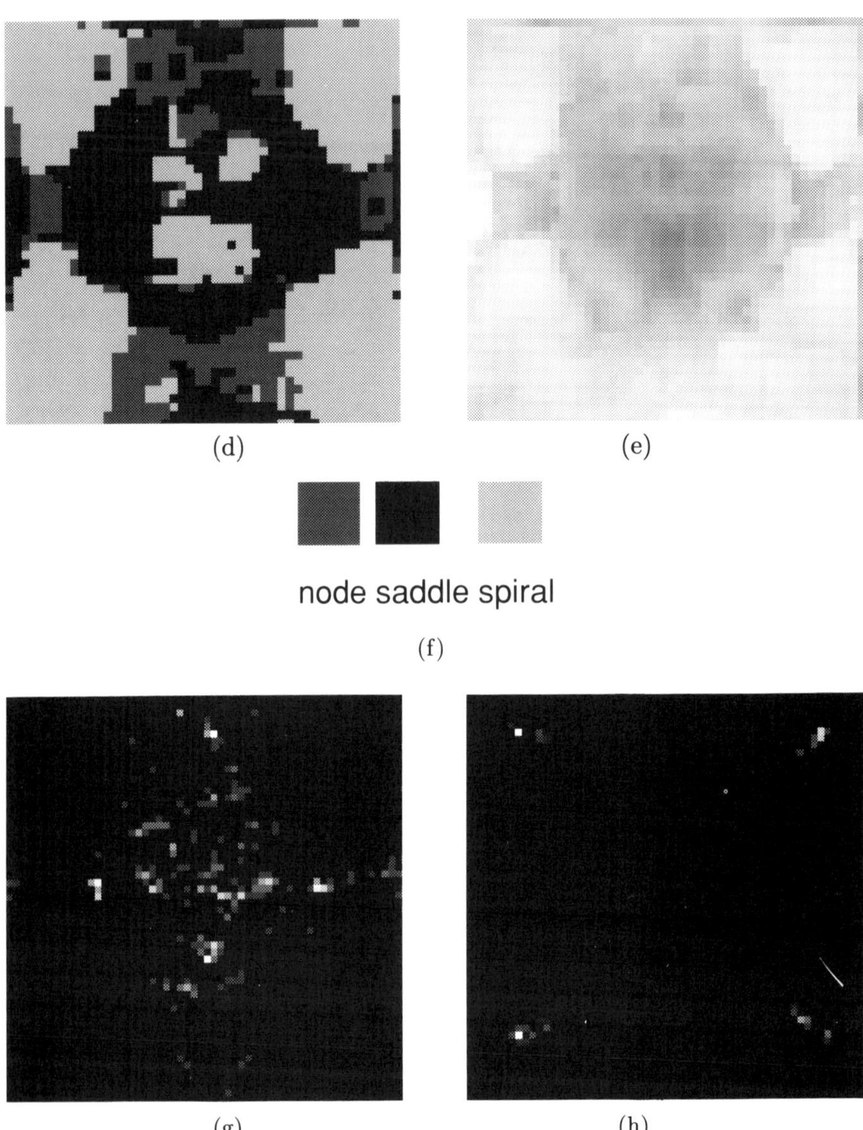

FIGURE 3.16. (d) The segmented image. Note that the areas close to the four corners are classified as spirals. Examination of the original picture shows that these areas are elliptical, which is a special case of spirals. Similarly, the areas around the central cylinder are classified as saddles. Again, this makes sense because these areas seem like sections of saddles. (e) The closeness of fit C_W, as defined in equation 3.22. (f) Shading key for phase portraits. (g) Map of saddle fixed points. The saddle fixed points are at expected locations. (h) Map of spiral fixed points. Observe that the locations of the spiral fixed points are towards the four corners of the image, as one would expect.

```
Description of given image:
  Spiral fixed points located
  at following points:-
  (5, 7)     (5, 51)
  (46, 48)  (48, 7)

Values of A and B matrices are:-

A = 0            1
    -1.031    0.322
B = -0.563   -0.496   at (5, 7)


A = 0            1
    -0.854   -0.276
B = -0.127    0.416  at (5, 51)


A = 1           -1.345
    1.292    -0.511
B = 3.737    -0.464  at (46, 48)


A = 1            1.517
    -1.774   -0.924
B =-0.959     1.056   at (48, 52)
```

FIGURE 3.17. (a) A symbolic description corresponding to the original image in figure 3.16 that is automatically generated by the program.   (b) The reconstructed flow image, based on the preceding symbolic description. Observe that the salient feature of the original image, namely the spiral-like nature of the texture has been captured in the reconstructed image. Furthermore, the orientation of the spirals is towards the center of the image.

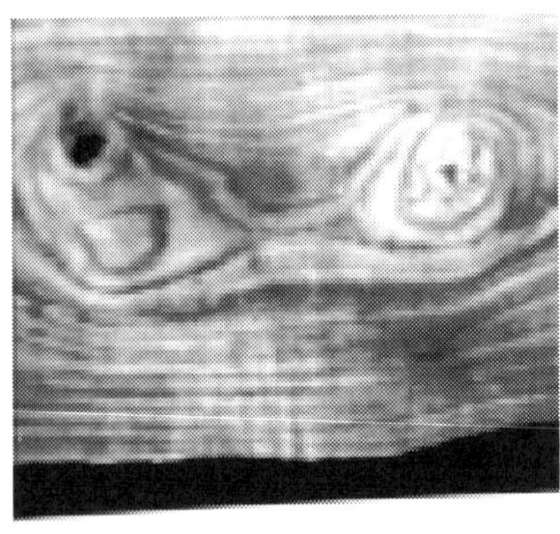 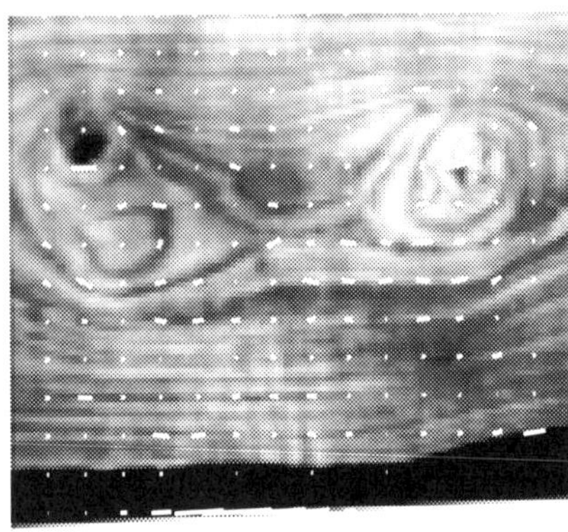

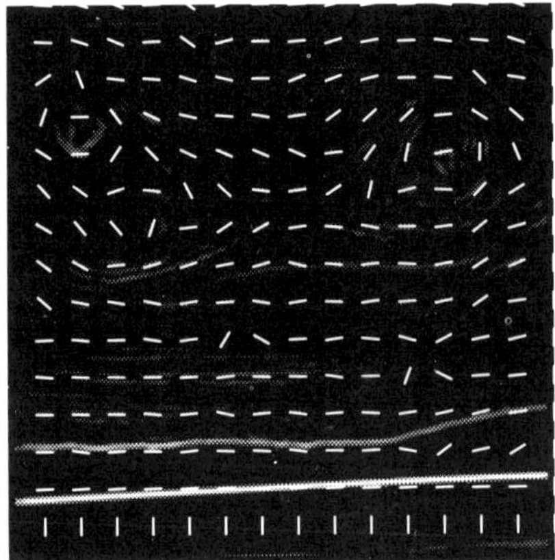

FIGURE 3.18. (a): The original flow image (b) The orientation field overlayed on the original image. The length of each oriented segment is proportional to the coherence. (c) The coherence image, with the orientation field overlayed.

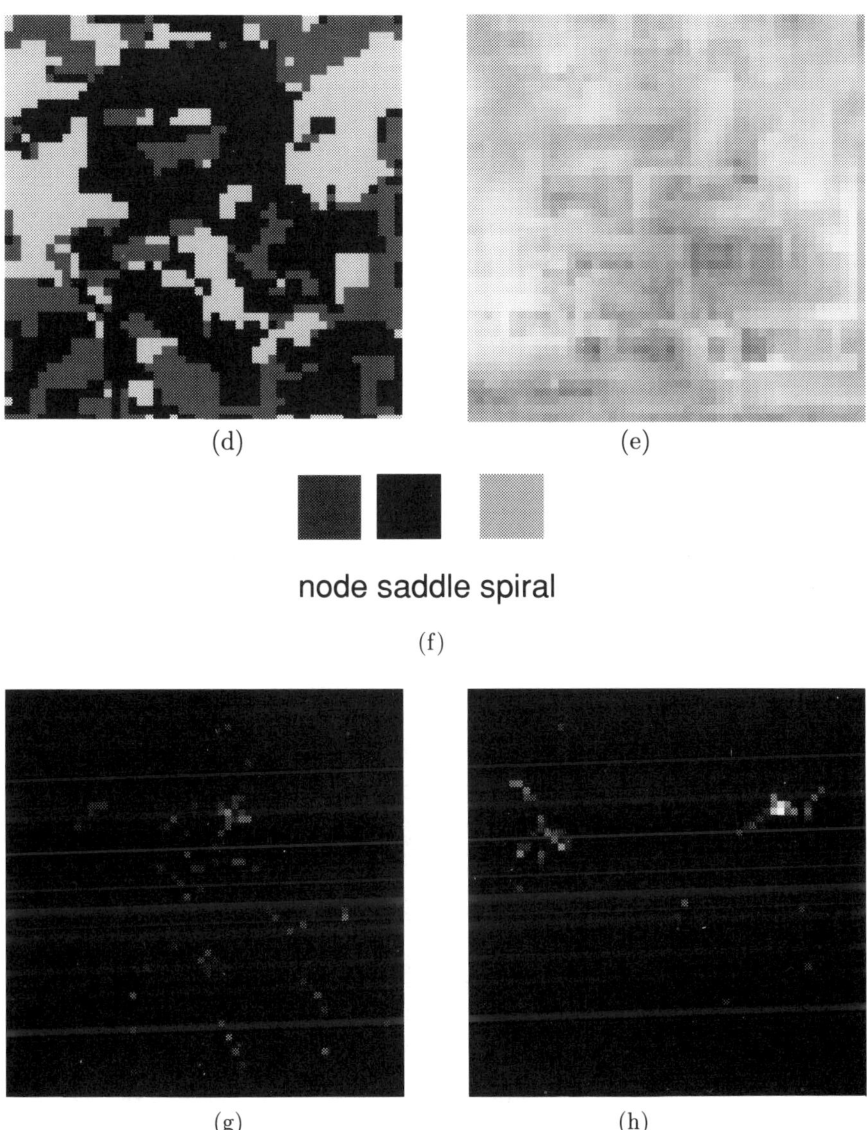

FIGURE 3.18. (d) The segmented image. Note that the area around each knot is classified as spiral-like, and the area between the knots is classified as saddle-like. (e) The closeness of fit C_W, as defined in equation 3.22. (f) Shading key for phase portraits. (g) Map of saddle fixed points (h) Map of spiral fixed points

```
Description of given image:
   Saddle fixed points located
   at following points:-
       (17, 32)

Values of A and B matrices are:-

A = 0             1
    0.839 -0.237
B = 1.313  -0.815 at (17, 32)


 Spiral fixed points located
   at following points:-
       (15, 44)  (20,  11)

A =  0             1
    -0.513  -0.336
B = -1.198    1.140 at (15, 44)


A =  0             1
    -0.536   0.344
B =  0.599  -0.020 at (20, 11)
```

FIGURE 3.19. (a)The symbolic description corresponding to the original image in figure 3.18, which is automatically generated by the program. (b): The reconstructed flow image based on the preceding symbolic description. Observe that the salient feature of the original image, namely the two knots, and the saddle in the middle have been captured in the reconstructed image. Observe that the parameters associated with the symbolic description (the **A** and **b** matrices) allow us to reconstruct a true likeness of the original texture.

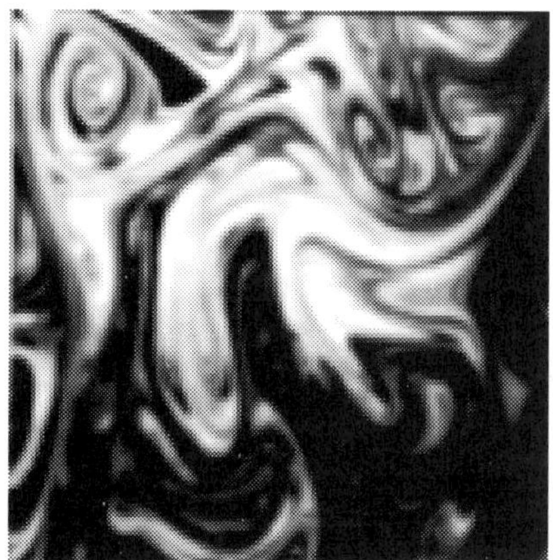 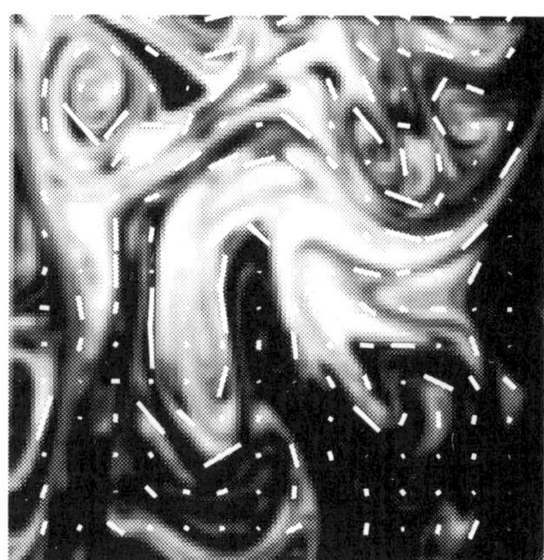

FIGURE 3.20. (a): The original flow image (b) The orientation field overlayed on the original image. The length of each oriented segment is proportional to the coherence. (c) The coherence image, with the orientation field overlayed.

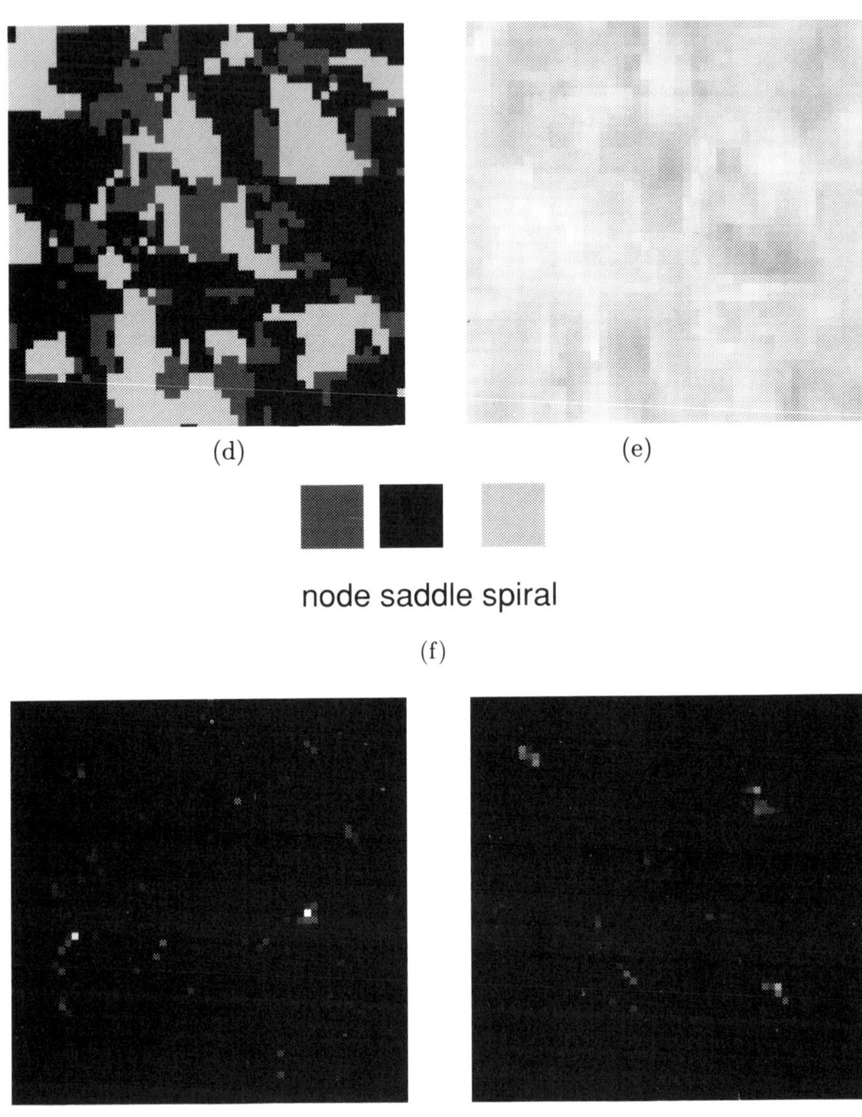

FIGURE 3.20. (d) The segmented image. The different types of phase portraits possible have been coded. (e) The closeness of fit C_W, as defined in equation 3.22. (f) Shading key for phase portraits. (g) Map of saddle fixed points (h) Map of spiral fixed points

```
 Description of given image:
    Node fixed points located
    at following points:-
         (16, 19)
Values of A and B matrices are:-
A =  0        1
    -0.207  -1.159
B =  0.727  -1.020 at (16, 19)


Saddle fixed points located
at following points:-
    (30, 43)    (33, 9)
Values of A and B matrices are:
A = 0        1
    0.889    0.25421
B = 0.613  -0.730459 at (30, 43)


A = 0          1
    0.118 -0.0545431
B =  0.528  0.0371258 at (33, 9)


Spiral fixed points located
 at following points:-
   (7, 7) (13, 41) (41, 44)
A =  0        1
    -0.508  -0.132
B = -1.231    0.325 at (7, 7)


A =  0       1
    -0.274  0.143
B = 0.623  -0.113 at (13, 41)


A =  0       1
    -0.447  0.282
B = 0.600   0.369 at (41, 44)
```

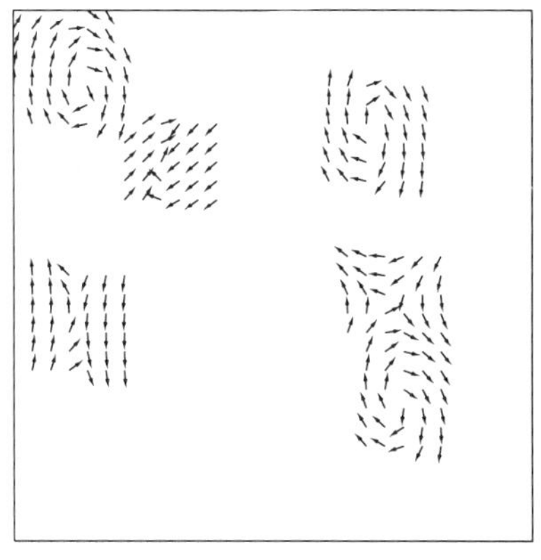

FIGURE 3.21. (a)The symbolic description corresponding to the original image in figure 3.20 which is automatically generated by the program. (b): The reconstructed flow image based on the preceding symbolic description. Observe that though the original image has a complex structure, the final symbolic representation can guide our interpretation of this image. In fact, if one analyzes the output carefully, one can see that the segmentation indeed makes sense when compared with the original image. Again, salient features have been captured in the reconstructed image. The parameters associated with the symbolic description (the **A** and **b** matrices) allow us to reconstruct a true likeness of the original texture.

3.5.5 IMAGE OF VORTEX FLOW

The visualization of vortex formation forms an important problem in experimental fluid mechanics. A typical problem is to observe how a fluid layer exhibits waves, rolls up into vortices and finally breaks down [107]. The tracking of these vortices as time evolves is of interest to researchers in experimental fluid mechanics. The method we have developed is capable of performing such an analysis, as indicated by preliminary experiments.

Figure 3.22 analyzes vortices behind a rotating propellor [118, p. 42]. We have run our algorithm to isolate the locations of vortices, and the intermediate results are presented in figures 3.22(b) through (g).

Reconstruction of vortex flow

Figure 3.23 shows the symbolic description of the image in figure 3.22 and its reconstruction based on values for $\mathbf{A}$ and $\mathbf{b}$ at the saddle and spiral fixed points.

An interesting aspect of such an analysis of fluid motion is that it allows automatic measurement of several parameters defined in experimental fluid mechanics. We give a concrete example, taken from [93]. A typical description of vortex flow [93], is as follows:

> ".. When the vortex street is breaking down, the pitch of the vortices does not change, but the width of the vortex street decreases."

The *width* of a vortex street is defined as the horizontal distance between centers of vortices, and the *pitch* of the vortices is defined as the vertical distance between centers of vortices. Based on the results obtained in figure 3.23, we can easily provide quantitative measurements for the width and pitch of the given flow. Thus, the method that we have proposed is not only capable of producing qualitative descriptions of a texture, but also capable of making quantitative measurements.

3.5.6 ANALYSIS OF A RESIST GEL DEFECT

In section 1.4.2, chapter 1 we considered the description of an image showing a resist gel defect. Figure 3.24a shows a 240x240 pixel image of resist gel defect, obtained from [32]. The result of applying the orientation estimation algorithm described in chapter 2 is displayed in figure 3.24b. The orientation field is calculated for each point in the image, but is displayed in a sampled form in order to avoid clutter.

In (figure 3.24b), the orientation field is overlayed on the original image to aid comparison. Figure 3.24c shows the initially segmented flow image. The segmented regions are gray-level coded, according to the scheme shown in figure 3.24d.

Figure 3.24e shows the spiral fixed point map. The location of the fixed point agrees with the nature of the original defect image.

FIGURE 3.22. (a): The original flow image *(Photograph by F. N. M. Brown, courtesy of University of Notre Dame)*. (b) The orientation field overlayed on the original image. The length of each oriented segment is proportional to the coherence. (c) The coherence image, with the orientation field overlayed.

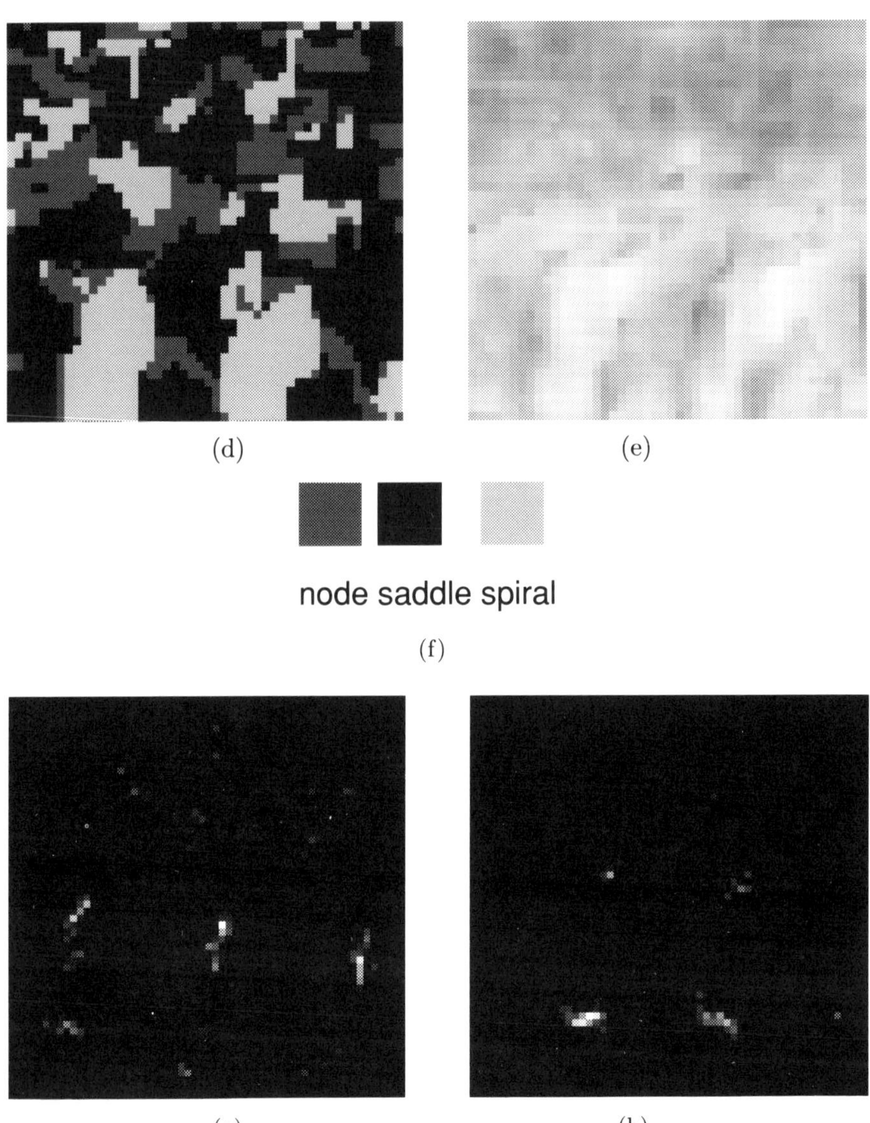

FIGURE 3.22. (d) The segmented image. The results of the segmentation show that the areas around the vortices are classified as spirals. (e) The closeness of fit C_W, as defined in equation 3.22. (f) Shading key for phase portraits. (g) Map of saddle fixed points. (h) Map of spiral fixed points. Observe that the locations of the spiral fixed points matches the actual locations of vortices in the original image.

```
Description of given image:
   Saddle fixed points located
   at following points:-
      (29, 11)  (30, 10)  (35, 29)
      (38, 51)  (46, 8)

A = 1          0.439
    -1.528 -1.624
B = -0.737 -0.602 at (29, 11)
A = 1          0.524
    -1.463 -1.849
B = -0.599  -0.332 at (30, 10)
A = 1          0.172
    -0.898  -1.454
B = -0.266  2.453   at (35, 29)
A = 1          0.039
    -0.147  -1.514
B = 0.580  1.143    at (38, 51)
A = 0            1
    0.688   0.269
B = 0.235  0.506    at (46, 8)

  Spiral fixed points located
  at following points:-
      (25, 20)  (27, 39)
      (45, 16)   (45, 34)

A = 0            1
    -1.116  -0.803
B = 1.144  -0.430   at (25, 20)
A = 0            1
    -1.324  1.217
B = 1.425  -2.457    at (27, 39)
A = 0          1
    -1.144  0.116
B = 0.112  0.195    at (45, 16)
A = 0          1
    -0.945  0.716
B = -0.246  -0.472 at (45, 34)
```

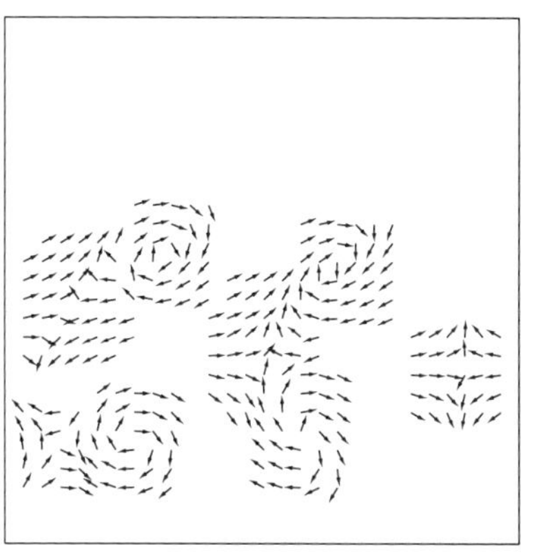

FIGURE 3.23. (a)The symbolic description corresponding to the original image in figure 3.22, which is automatically generated by the program. (b): The reconstructed flow image based on the preceding symbolic description. The algorithm is able to correctly identify the vortices present in the given image. Furthermore, the reconstruction bears a close resemblance to the original textured image.

Reconstruction of the resist gel defect

The symbolic description of the image is automatically generated from the program as shown in figure 3.25(a). Figure 3.25(b) shows the reconstructed flow image based on values for **A** and **b** at the saddle fixed points.

3.5.7 ANALYSIS OF TEXTURED PAINT BRUSH STROKES

Let us consider an example from a totally different domain, and move into the world of paintings. Suppose we were interested in analyzing a painting by examining the nature of brush strokes. We have chosen to analyze a painting by Van Gogh in this manner. The reason for this choice is that many of Van Gogh's paintings are characterized by vigorous brush strokes, that lend an interesting flow-like nature to the paintings. This example illustrates the power of using the phase portrait model for texture description and interpretation in diverse domains.

Figure 3.26 analyzes a painting by Van Gogh [74], entitled '*A starry night*'.

Reconstruction of textured paint brush strokes

The Van Gogh painting provides a very interesting case to analyze. Since the painting is dominated by rich textural pattern, the reader is urged to compare the reconstruction results very carefully with the original painting. The reconstruction results may appear to be very cluttered, but upon careful examination with respect to the original painting, they seem to make perfect sense. This has been the experience of the authors while interpreting the results of reconstruction.

Figure 3.27 shows the reconstructed flow image based on values for **A** and **b** at the saddle and spiral fixed points. During this reconstruction, the threshold mentioned in 3.4.6 was set at 50%, as in earlier examples. This means that we are concentrating on the dominant features of the given texture.

Let us see what happens when we allow the threshold to be relaxed, thus admitting more features. The threshold was relaxed up to 30%, resulting in a description shown in figure 3.28. This corresponds to a finer, more detailed description as we allow more features to be present.

This example serves to illustrate the role of the strength of a stimulus in perceptual tasks. As we start focussing on less dominant features of a scene, a more detailed description of the scene begins to emerge. The threshold described in section 3.4.6 is just a beginning in this direction. More complex and elaborate schemes can be used to control such a focussing mechanism. We have used a simple, global threshold. This could be replaced with an adaptive scheme, that adjusts its behaviour over the entire image. We plan to pursue this direction of research in the future.

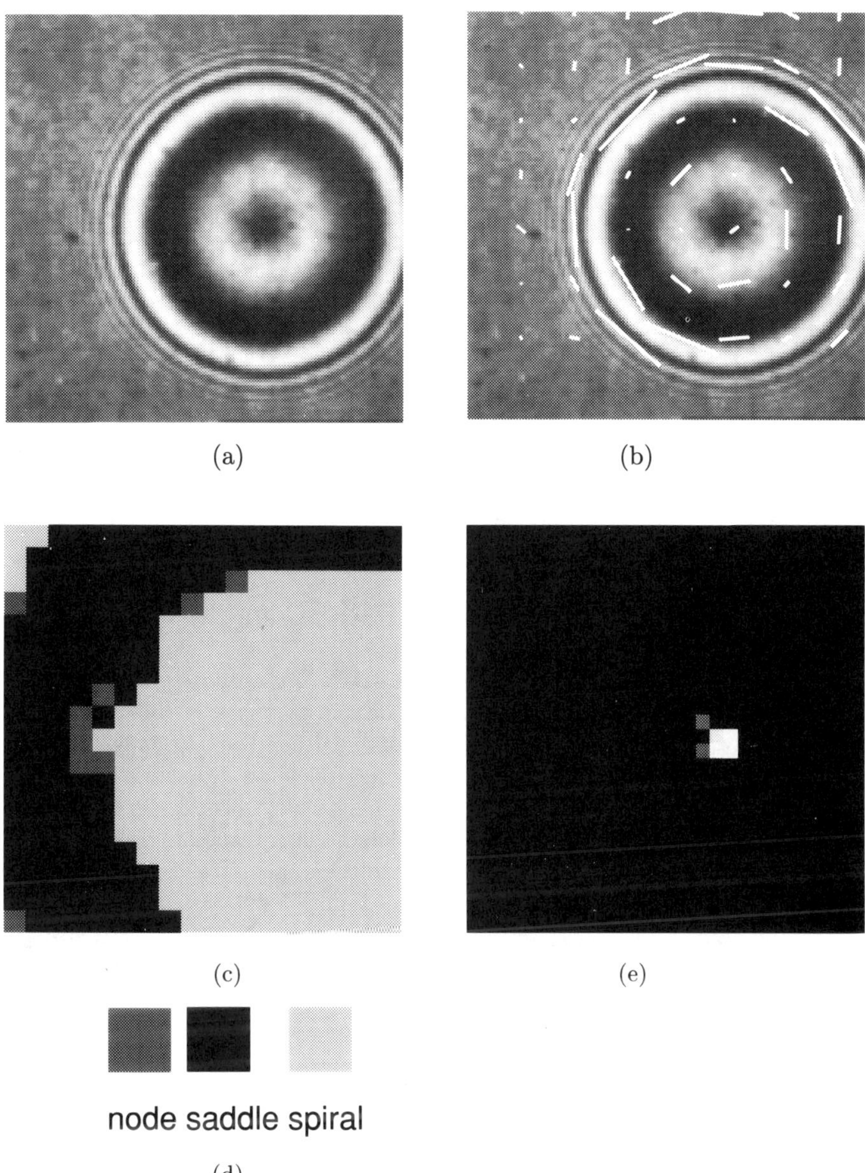

FIGURE 3.24. (a): The original image of the resist gel defect (Reproduced with permission from *Integrated Circuit Mask Technology* by D. J. Elliott, ©1985 Mc-Graw Hill). (b) The orientation field overlayed on the original image in the form of white segments. The length of each oriented segment is proportional to the coherence. (c) The segmented image. The different types of phase portraits possible have been coded. Note that the areas around the defect are classified as spiral regions. (d) Shading key for phase portraits. (e) Map of spiral fixed points

```
Description of given image:
  Spiral fixed points located
  at following points:-
  Row = 14, col = 17;

  Values of A and B matrices are:
  A  =  0          1
        -1.0402   -0.0017

  B  =  -0.3111  0.4165
        at  (14, 17)
```

FIGURE 3.25. (a): This is the symbolic description of the input image. (b): The reconstructed image, based on this symbolic description. This corresponds to the original image in figure 3.24 Observe that the salient feature of the original image, namely the concentric circles of the texture has been captured in the reconstructed image.

3.6   Experiments with noise addition

In this section we briefly explore the robustness of the phase portrait technique with respect to noise addition. One of the expected benefits of a qualitative description is the robustness of the description to image degradation. This expectation is borne out, as the following experiments show.

Images were degraded by adding random Gaussian noise. Specifically, we used the routine **GASDEV** from [100] to add zero mean Gaussian noise with a specified standard deviation to each pixel in the image. The resulting value was hardlimited to lie in the [0,255] range. Figure 3.29(a) shows the result of adding zero mean Gaussian with a standard deviation of 10 gray levels to the original image. All the parameters used for this figure are the same as in figure 3.16. The segmentation shown in figure 3.29(c) is very similar to the one in figure 3.16(d). The salient features of the orientation field as shown in figure 3.29(d) are also nearly the same as figure 3.17.

Figure 3.30(a) shows the result of adding zero mean Gaussian with a standard deviation of 20 gray levels. Though the qualitative nature of figure 3.30(c) has not changed much, the algorithm does not regard the spiral at the top left corner to be salient any longer.

In figure 3.31(a), the standard deviation of the noise was increased to 40 gray levels. The results show that the algorithm performs poorly, and one may be led to believe that this is the performance limit. However, the orientation field extracted in figure 3.31(a) still retains a strong resemblance to the original orientation field. There are two ways to explain this behaviour. First, as increasing amounts of noise are added, the size of the region over which linearity can be assumed decreases. Second, least squares techniques

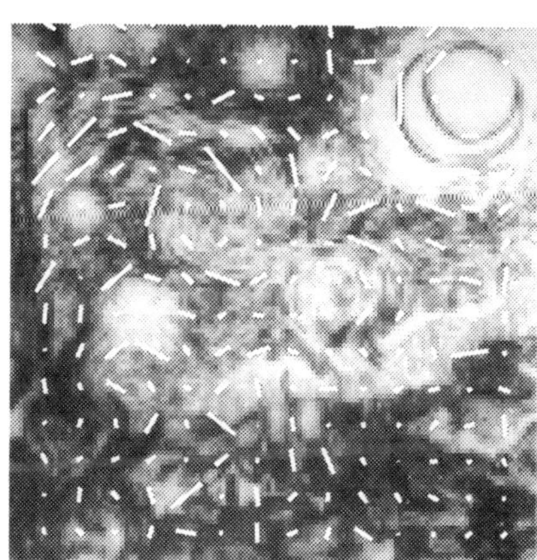

FIGURE 3.26. (a): The original painting (Reproduced with permission from Editions d'Art Albert Skira). (b) The orientation field overlayed on the original image. The length of each oriented segment is proportional to the coherence. (c) The coherence image, with the orientation field overlayed.

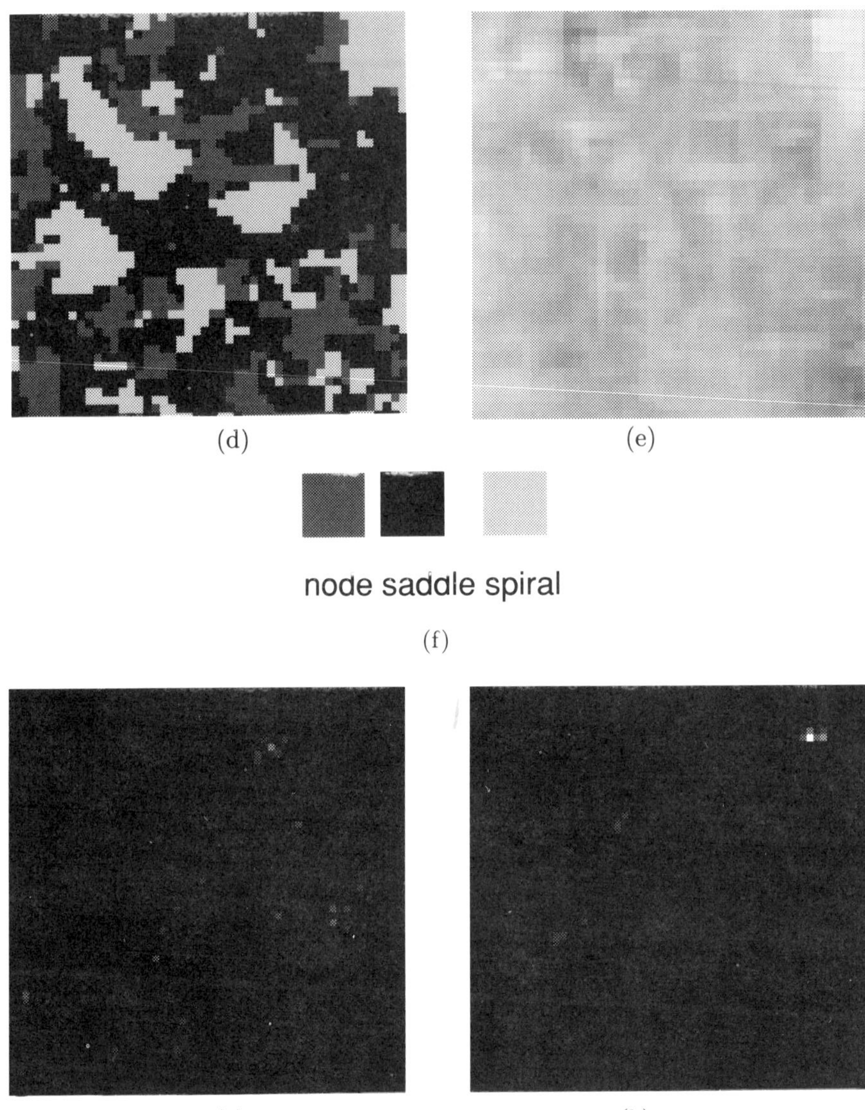

FIGURE 3.26. (d) The segmented image. The different types of phase portraits possible have been coded. (e) The closeness of fit C_W, as defined in equation 3.22. (f) Shading key for phase portraits. (g) Map of saddle fixed points (h) Map of spiral fixed points

```
Description of given image:
 Saddle fixed points located
 at following points:-
      (8, 38)


A = 0         1
    0.297  0.081
B = 0.972  -0.105 at (7, 38)


 Spiral fixed points located
 at following points:-
             (7, 49)


A = 0          1
   -0.882  -0.008
B = -0.791  -0.576 at (7, 49)
```

FIGURE 3.27. (a) The symbolic description at a coarse level (corresponding to the original image in figure 3.26) (b): The reconstructed flow image at a coarse level of description. The most dominant features of the image at this level happen to be the saddle and spiral at the upper right corner. This represents the area around the moon in the original painting. This makes sense, because the moon is indeed the dominant element in the painting.

```
Description of given image:
 Node fixed points located
 at following points:-
 (8, 20)  (23, 31)  (49, 5)

 Saddle fixed points located
 at following points:-
  (8, 38)  (19, 42)  (31, 47)
 (32, 39) (33, 47)
 (38, 21) (43, 2)

 Spiral fixed points located
 at following points:-
   (6, 48)
```

FIGURE 3.28. (a) The symbolic description generated at a finer level (corresponding to the original image in figure 3.26). We have omitted the elements of the **A** and **b** matrices. (b) Reconstruction based on the preceding symbolic description, and performed at a finer level of description. One can see that many more features are present in the reconstructed image, as compared to figure 3.26. The reconstructed image is rich in the number of saddles, and this agrees with the original image. However, there are not as many spiral fixed points as one would have expected. This is perhaps due to the fact that the scale at which the non-linear least squares fit was performed is not attuned to the feature sizes in the original image.

are well known to be sensitive to noise. Hence the algorithm we have suggested should be modified to deal with outliers in the spirit of techniques from robust statistics. Since the latter involves a change in the algorithm, we simply modified the size of the window used for least squares fitting. Instead of using a window size of 11x11 to perform a least squares fit, we used a window of size 7x7, which represents a reduction of half in the area of fit.

The resulting segmentation, shown in figure 3.32(a) is again similar to figure 3.16(d). However, the reconstructed orientation field shows that the spirals in the corners are no longer the only dominant features.

Finally, figure 3.33(a) was generated with Gaussian noise having a standard deviation of 60 gray levels. Though the resulting image is considerably degraded, the orientation field still captures the flow-like nature of the original image. The segmentation still bears a resemblance to the noise-free case. The salient features captured in the reconstructed field in 3.33(d) are faithful to the orientation field in 3.33(b). Though spirals are detected at the four corners, the saddles at the image center have also become important.

In conclusion, these experiments indicate that the phase portraits technique exhibits graceful degradation in the presence of noise.

3.7   A related model from fluid flow analysis

It is very interesting to observe that a similar classification scheme for flows has been devised by researchers in the area of fluid dynamics. Though such techniques are well known (see Petterssen [98] for instance), they have been recently revisited (see Ottino[94]). We briefly summarize the treatment of velocity field properties, as presented in [98, pp.32-43]. Popular notation in fluid dynamics uses u and v to refer to velocity components in the two dimensional case. These are regarded as functions of x and y, viz.,

$$u = u(x, y) \quad v = v(x, y) \tag{3.24}$$

Let us use a Taylor series expansion for u and v about the origin as follows

$$u = u_0 + \left(\frac{\partial u}{\partial x}\right)_0 x + \left(\frac{\partial u}{\partial y}\right)_0 y + Higher\ order\ terms$$

$$v = v_0 + \left(\frac{\partial v}{\partial x}\right)_0 x + \left(\frac{\partial v}{\partial y}\right)_0 y + Higher\ order\ terms \tag{3.25}$$

where u_0 and v_0 are the velocity components at the arbitrarily chosen origin. The higher order terms may be dropped if their relative contribution is negligible compared to the linear terms. One can rewrite equation 3.25

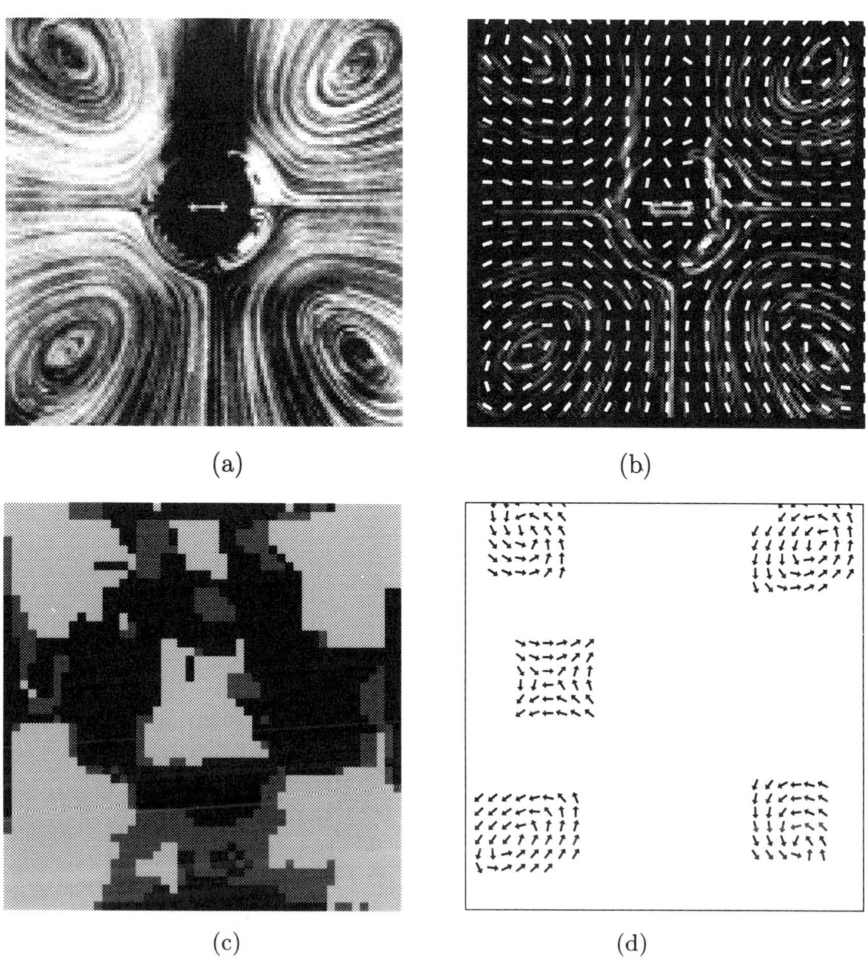

FIGURE 3.29. (a): The flow image in figure 3.16 after adding Guassian noise with zero mean and standard deviation of 10 gray levels. (b) The orientation field overlayed on the coherence image. Unit vectors are used for the orientation field. (c) The result of segmenting the above image. A window of 11x11 pixels was used to perform the non-linear least squares fit. (d) The reconstructed orientation field.

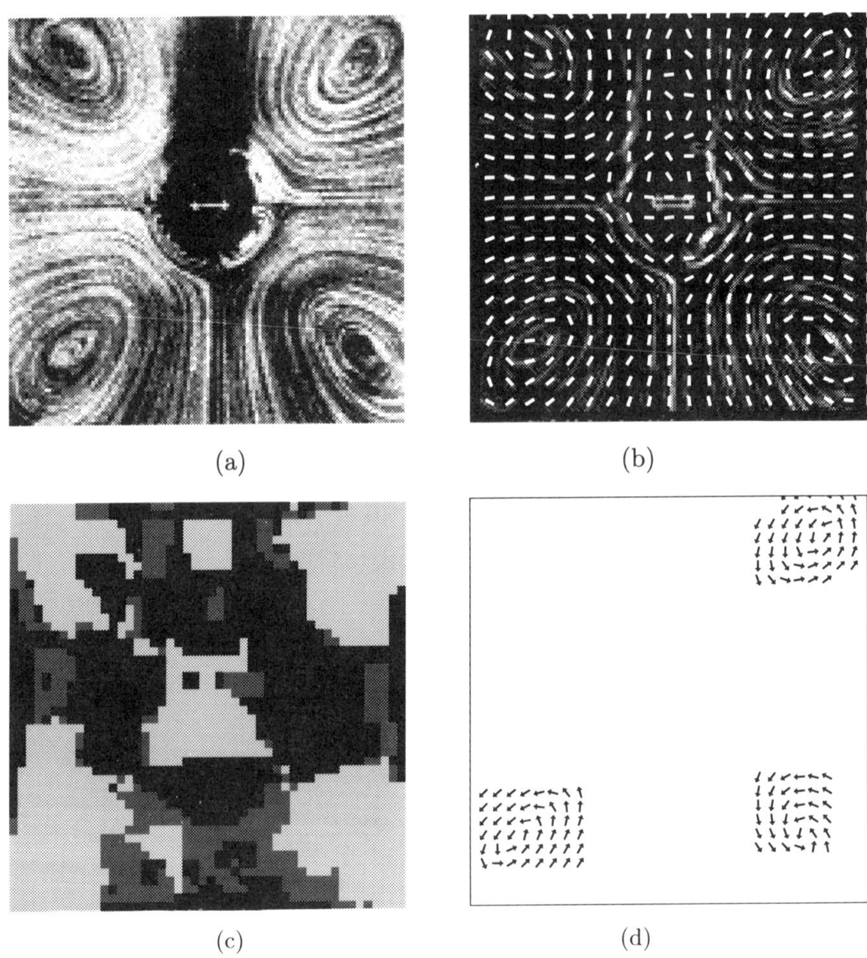

(a)   (b)

(c)   (d)

FIGURE 3.30. (a): The flow image in figure 3.16 after adding Guassian noise with zero mean and standard deviation of 20 gray levels. (b) The orientation field overlayed on the coherence image. Unit vectors are used for the orientation field. (c) The result of segmenting the above image. A window of 11x11 pixels was used to perform the non-linear least squares fit. (d) The reconstructed orientation field.

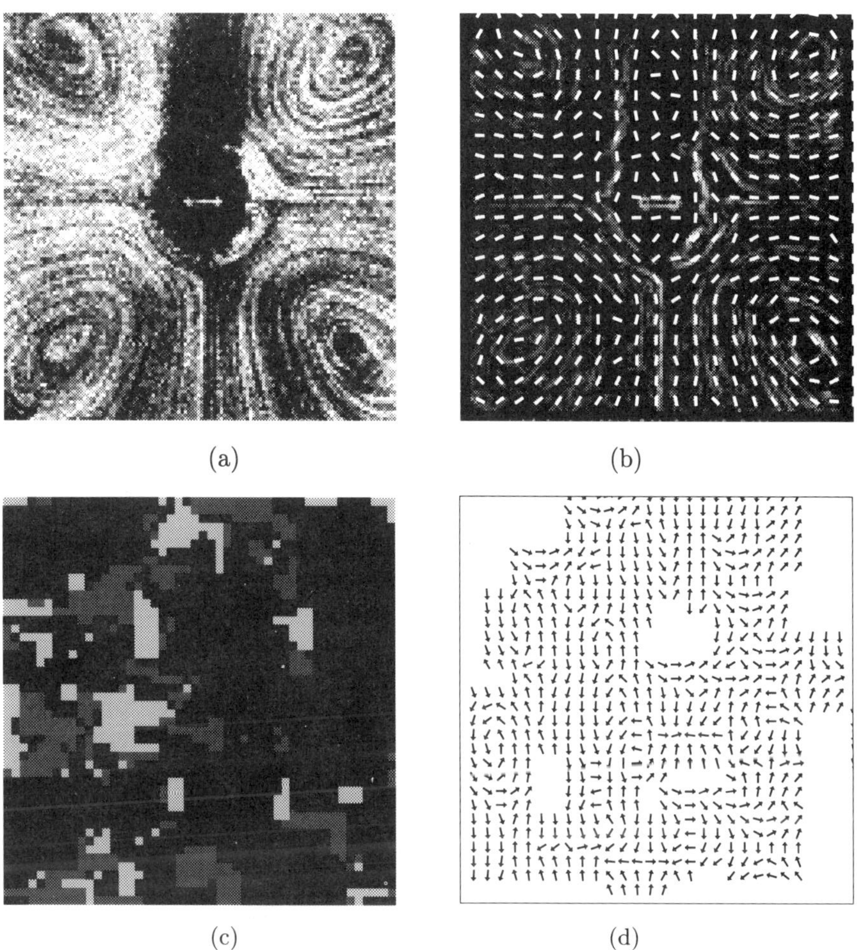

(a)   (b)

(c)   (d)

FIGURE 3.31. (a): The flow image in figure 3.16 after adding Guassian noise with zero mean and standard deviation of 40 gray levels. (b) The orientation field overlayed on the coherence image. Unit vectors are used for the orientation field. (c) The result of segmenting the above image. A window of 11x11 pixels was used to perform the non-linear least squares fit. (d) The reconstructed orientation field.

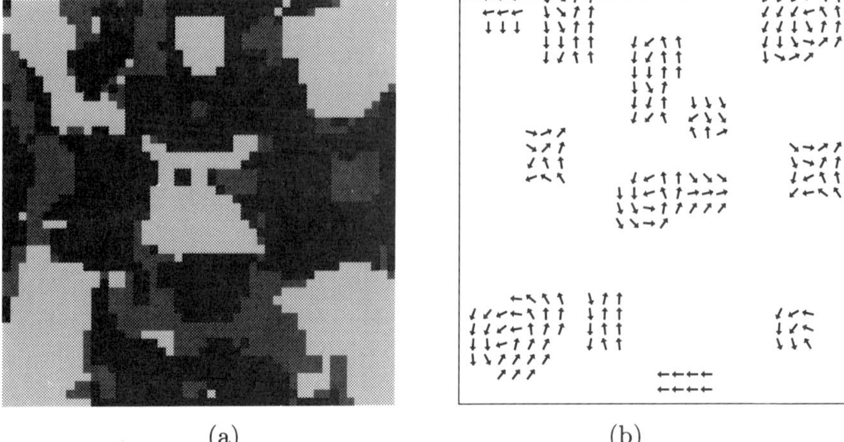

(a)   (b)

FIGURE 3.32. (a) The result of segmenting the image in figure3.31(a). Note that a window of 7x7 pixels was used to perform the non-linear least squares fit. (b) The reconstructed orientation field.

using the substitutions

$$D = \left(\frac{\partial u}{\partial x} + \frac{\partial v}{\partial y}\right)_0 \qquad F = \left(\frac{\partial u}{\partial x} - \frac{\partial v}{\partial y}\right)_0$$

$$q = \left(\frac{\partial v}{\partial x} - \frac{\partial u}{\partial y}\right)_0 \qquad r = \left(\frac{\partial v}{\partial x} + \frac{\partial u}{\partial y}\right)_0 \qquad (3.26)$$

to yield the following

$$u = u_0 + \frac{1}{2}(D + F)x + \frac{1}{2}(r - q)y$$

$$v = v_0 + \frac{1}{2}(r + q)x + \frac{1}{2}(D - F)y \qquad (3.27)$$

The coefficients D and q are independent of choice of axes, whereas F and r are not, as shown in appendix A. D and q are the divergence and curl of the two dimensional velocity field. If we rotate the axes through an angle θ_0 such that $\partial v/\partial x = -\partial u/\partial y$, then r becomes 0, and equation 3.27 can be rewritten as

$$u = u_0 + \frac{1}{2}Fx + \frac{1}{2}Dx - \frac{1}{2}qy$$

$$v = v_0 - \frac{1}{2}Fy + \frac{1}{2}Dy + \frac{1}{2}qx \qquad (3.28)$$

These equations refer to a *linear field of motion* [98], and are valid only when the region of analysis is sufficiently small to justify negligence of

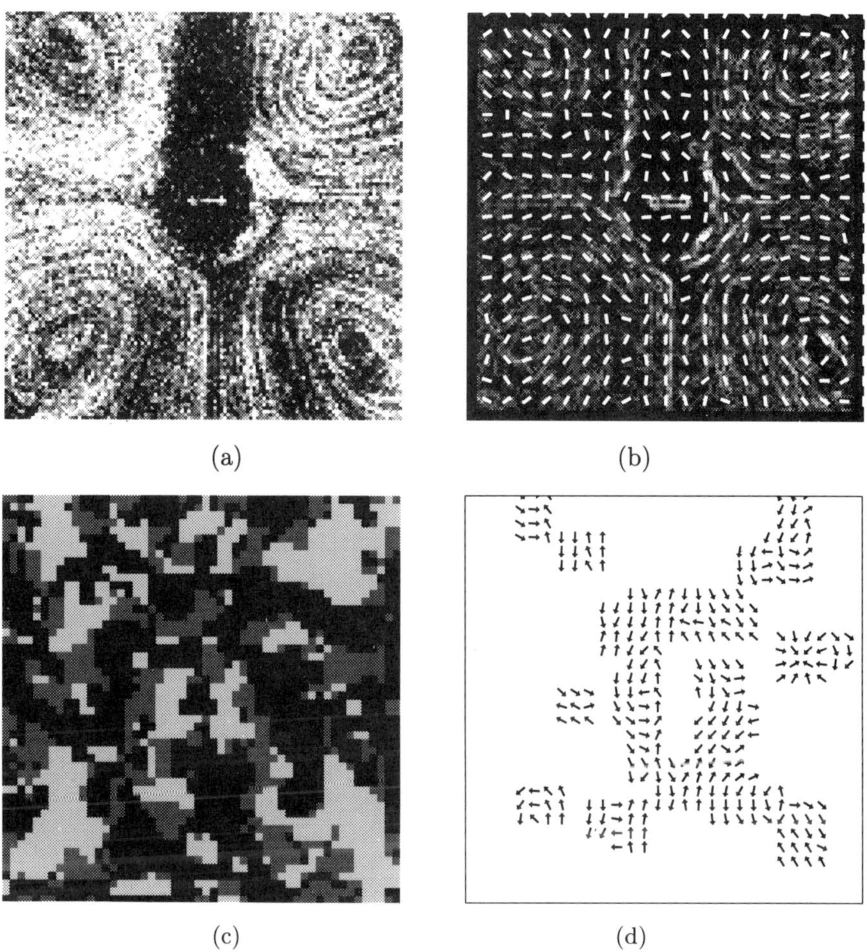

(a)   (b)

(c)   (d)

FIGURE 3.33. (a): The flow image in figure 3.16 after adding Guassian noise with zero mean and standard deviation of 60 gray levels. (b) The orientation field overlayed on the coherence image. Unit vectors are used for the orientation field. (c) The result of segmenting the above image. A window of 7x7 pixels was used to perform the non-linear least squares fit. (d) The reconstructed orientation field.

Interpretation of motion components for a velocity field		
Term	Definition	Interpretation
u_0, v_0	u_0, v_0	Translation
D	$u_x + v_y$	Expansion or divergence
F	$u_x - v_y$	Deformation
q	$v_x - u_y$	Rotation or vorticity

TABLE 3.1. Motion components

higher order terms in equation 3.25. An interesting kinematical interpretation can be attached to the terms in equation 3.28. The basic idea behind the interpretation is that a two-dimensional sheet of fluid can be brought from/position to another by a *translation*, a *deformation*, an *expansion* and a *rotation*. The mathematical definition for these terms are presented in table 3.1, with the subscript notation used to represent partials.

The physical significance of these component motions is as follows:

1. *Translation.* The terms u_0 and v_0 represent a uniform translation. A region of fluid taking part in this motion will not undergo any change in shape.

2. *Expansion.* The divergence D of a two dimensional flow is the areal expansion per unit area per unit time, and is independent of the translation, deformation and rotation.

3. *Deformation.* The term F represents the rate at which a unit square would be transformed into a rectangle of the same size, such that the area is stretched in the direction of the axis of dilation and shrunk in the direction of contraction.

4. *Rotation.* The term q constitutes a rotation about the origin. Thus, any region will be rotated about the origin without changing its shape or size.

If one were to plot the distribution of velocity components for each of the above cases, one would end up with figure 3.34.

3.7.1 A COMPARISON BETWEEN THE FLUID MOTION VIEWPOINT AND THE PHASE PORTRAIT VIEWPOINT

Though the material in this section is not new (see Ottino[94] for a similar discussion), it is presented here for the sake of completeness.

Let us change the notation in order to conform to that in section 3.3.1 and replace x and y by x_1 and x_2 respectively. Also, noting that $u =$

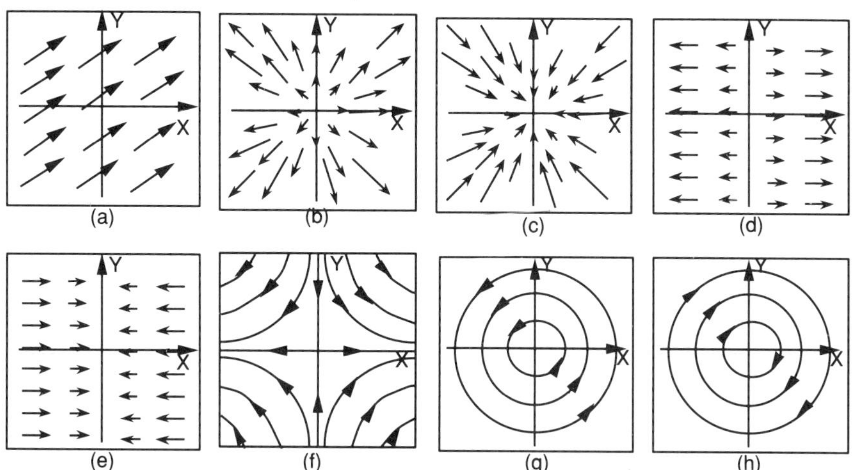

FIGURE 3.34. The velocity components of a linear field. (a) Uniform translation (b) Positive divergence, or areal expansion (c) Negative divergence, or areal contraction (d) The x-component of deformation, or the dilation. (e) The y-component of deformation, or the contraction (f) Streamlines for D, a combination of dilation and contraction. (g) Positive rotation (h) Negative rotation

$dx/dt = \dot{x} = \dot{x}_1$ and $v = dy/dt = \dot{y} = \dot{x}_2$, we replace u by $\dot{x}_1$ and v by $\dot{x}_2$. Furthermore, we can use the following substitutions

$$b_1 = u_0, \; a_{11} = \left(\frac{\partial u}{\partial x}\right)_0, \; a_{12} = \left(\frac{\partial u}{\partial y}\right)_0$$

$$b_2 = v_0, \; a_{21} = \left(\frac{\partial v}{\partial x}\right)_0, \; a_{22} = \left(\frac{\partial v}{\partial y}\right)_0 \tag{3.29}$$

so that using equation 3.26 we get

$$\mathbf{A} = \begin{bmatrix} a_{11} & a_{12} \\ a_{21} & a_{22} \end{bmatrix} = \frac{1}{2}\begin{bmatrix} D+F & r-q \\ r+q & D-F \end{bmatrix} \tag{3.30}$$

This is possible because the partial derivatives evaluated at the origin are constants. We now get back the familiar equation $\dot{x}(t) = \mathbf{X}(x) = \mathbf{A}x + \mathbf{b}$ discussed in section 3.3.2.3, with the above substitutions.

The advantage of viewing the system of equations $\dot{x}(t) = \mathbf{A}x + \mathbf{b}$ as representing fluid motion is that it allows for a *quantitative physical interpretation* of the various terms in the matrices $\mathbf{A}$ and $\mathbf{b}$. In terms of the matrix $\mathbf{A}$, we get

$$D = expansion\ or\ divergence = a_{11} + a_{22} = tr(\mathbf{A})$$

$$q = rotation\ or\ vorticity = a_{21} - a_{12} \div 2 \tag{3.31}$$

which are independent of rotation of axes, as shown in appendix A.

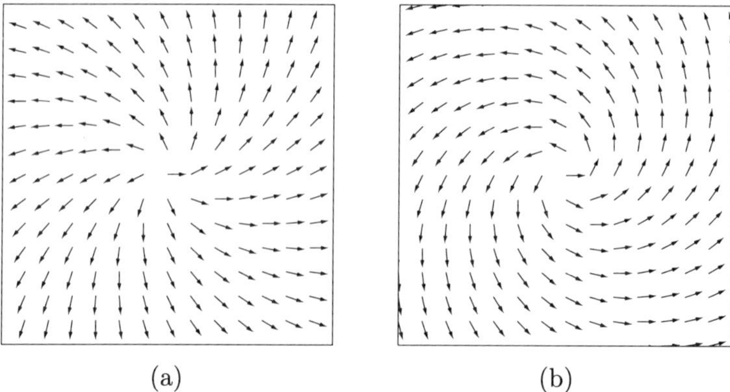

(a)   (b)

FIGURE 3.35. (a) A spiral with divergence dominating; The **A** matrix in this case was $a_{11} = 1$, $a_{12} = -0.5$, $a_{21} = 0.5$, $a_{22} = 1$. (b) A spiral with rotation dominating; The **A** matrix in this case was $a_{11} = 1$, $a_{12} = -2$, $a_{21} = 2$, $a_{22} = 1$.

Though the values of F and r depend on the choice of coordinate system, the quantity $\sqrt{F^2 + r^2}$ is invariant with respect to rotation of axes. The analysis leading to equation 3.28 was based on eliminating r by rotation, whereby F in the rotated system was the deformation. This is equivalent to considering $\sqrt{F^2 + r^2}$ in the original coordinate system as a measure of deformation. Thus we now have a physical interpretation for the terms of the matrices **A** and **b**, which is summarized in table 3.2. Such an interpretation is useful in analyzing combinations of expansion, deformation and rotation. We can determine which component, if any, dominates the local velocity (or orientation) field. Spiral flow (as depicted in figure 3.35 is actually composed of a combination of expansion (divergence) and rotation (vorticity), and it is now possible to determine which component is dominant. Thus, one could have descriptions such as a *strongly divergent spiral*, where the divergence dominates, or a *strongly rotational spiral*, where the rotational component dominates. This is illustrated in figure 3.35, which contrasts the two kinds of spirals.

In section 3.2.2 we had mentioned that wood grains can be characterized in terms of their curvature. The vorticity q and deformation D offer measures for the curvature in a vector field, which can be used to quantify descriptors used in wood inspection. In section 3.4 we shall present a technique to automatically estimate q and D in an intensity image.

3.7.2   CLASSIFICATION OF VELOCITY FIELDS

Once we have a local description of the velocity field, it can be classified according to the types of motion present. The following cases are of interest.

Interpretation of terms in A and b		
Matrix term	Equivalent fluid term	Interpretation
b_1, b_2	u_0, v_0	Translation
$a_{11} + a_{22}$	D	Expansion or divergence
$\sqrt{(a_{11} - a_{22})^2 + (a_{12} + a_{21})^2}$	F	Deformation
$a_{21} - a_{12}$	q	Rotation or vorticity

TABLE 3.2. Interpretation of matrix components

1. Existence of a kinematic center (or fixed point, or singular point)
   This is a point at which the velocity vanishes. Putting $u = v = 0$ in equation 3.28, we get the coordinates of the kinematic center (x_c, y_c) to be

$$
\begin{aligned}
x_c &= -2\frac{u_0(D - F) + v_0 q}{D^2 - F^2 + q^2} \\
y_c &= 2\frac{u_0 q + v_0(D + F)}{D^2 - F^2 + q^2}
\end{aligned}
\tag{3.32}
$$

If x_c and y_c are finite, there is a kinematic center; if either x_c or y_c or both are indeterminate, there is no center but a singular line along whcih the velocity is zero; if x_c and y_c are infinite, there is no kinematic center. The condition for the existence of a kinematic center is

$$
D^2 - F^2 + q^2 \neq 0
\tag{3.33}
$$

Recall that this condition holds in the rotated coordinate system. For the original coordinate system, this condition becomes $D^2 - (F^2 + r^2) + q^2 \neq 0$, which means the matrix $\mathbf{A}$, as in equation 3.30 is non-singular. Thus, in case $\mathbf{A}$ is singular, we could either have a singular line, or no singular point.

2. Existence of straight streamlines
   Assuming that a kinematic center exists, one can translate the coordinate system such that the terms u_0 and v_0 are zero at the origin, so that

$$
\begin{bmatrix} u \\ v \end{bmatrix} = \frac{1}{2} \begin{bmatrix} D + F & -q \\ q & D - F \end{bmatrix} \begin{bmatrix} x \\ y \end{bmatrix}
\tag{3.34}
$$

The condition for straight streamlines through the center is

$$
\frac{dy}{dx} = \frac{v}{u} = \frac{y}{x}
\tag{3.35}
$$

Combining this with equation 3.34, we get

$$tan\theta = \frac{F \pm \sqrt{F^2 - q^2}}{q} \tag{3.36}$$

where θ is the angle from the x-axis to the straight streamline. Table 3.3 summarizes the various cases possible.

3. Of the cases presented in table 3.3, the following deserve special attention. Consider the case

$$\begin{aligned}
q^2 - F^2 - D^2 &= 0 \\
\frac{u_0}{v_0} = \frac{q}{D+F} &= -\frac{D-F}{q}
\end{aligned} \tag{3.37}$$

so that equation 3.32 gives indeterminate values for x_c and y_c. In this case, all streamlines are straight, with a singular line along which the velocity vanishes.

Next, consider the case

$$\frac{u_0}{v_0} \neq \frac{q}{D+F} = -\frac{D-F}{q} \tag{3.38}$$

Here, we get curved streamlines without a kinematic center, and a great variety of such patterns is possible. Only a few have been shown in table 3.3.

3.7.3   IMPORTANCE OF DIVERGENCE AND CURL

Yet another way of viewing the computational scheme presented in this chapter is to regard the estimation of the matrices $\mathbf{A}$ and $\mathbf{b}$ as providing a method to find the curl and divergence of a noisy vector field. Specifically, the noise dealt with in this chapter is the rotation of arbitrarily chosen vectors through π radians. This connection is provided through equation 3.31.

There are very strong reasons for one to be interested in a computation of the vorticity and divergence of a vector field. This arises from the fact that one can go back and forth between a vector field on the one hand and its vorticity and divergence field on the other hand [9, p. 85]. The relation between a velocity field $\mathbf{v}$, its vorticity and divergence is given by

$$\begin{aligned}
vorticity &= \Omega = \nabla \times \mathbf{v} \\
divergence &= \Theta = \nabla \cdot \mathbf{v}
\end{aligned} \tag{3.39}$$

The inverse problem, that of determining the velocity from given distributions of the rotation and divergence is given by the following equations. For

Condition	Case
$F = q = 0$	Pure divergence
$F^2 > q^2$	Deformation predominates over rotation; streamlines are hyperbolic  Pure deformation — Deformation with positive rotation $(F > q)$ — Deformation with negative rotation $(F > -q)$
$q^2 > F^2$	Rotation predominates over deformation; no straight streamlines through the center 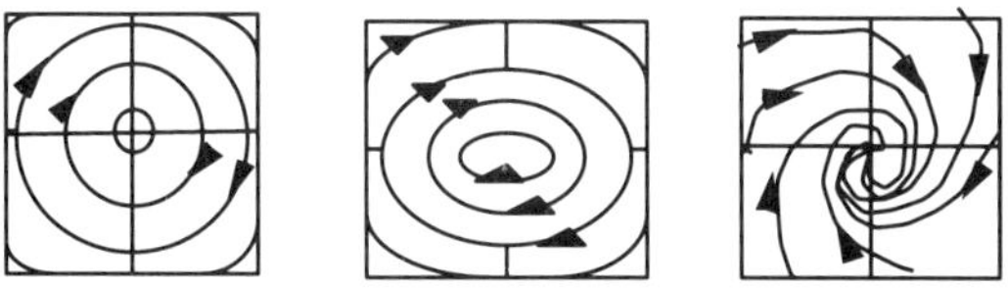 Pure rotation — Rotation and deformation — Rotation, deformation and convergence superimposed
$q^2 - F^2 = D^2$ and $\dfrac{v_0}{u_0} = \dfrac{q}{D+F} = -\dfrac{D-F}{q}$	All streamlines straight 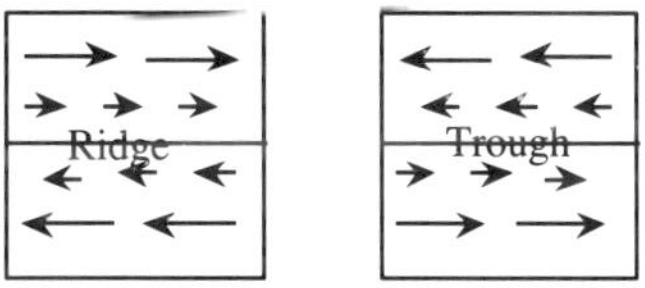 Ridge — Trough
$\dfrac{v_0}{u_0} \neq \dfrac{q}{D+F} = -\dfrac{D-F}{q}$	Curved streamlines without a center 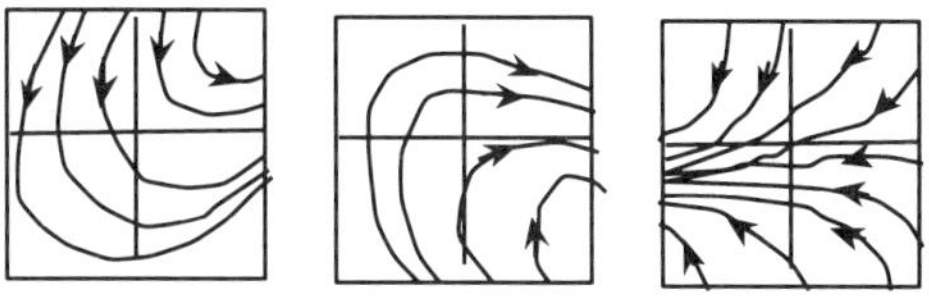

TABLE 3.3. Summary of various types of streamlines possible from equation 3.36

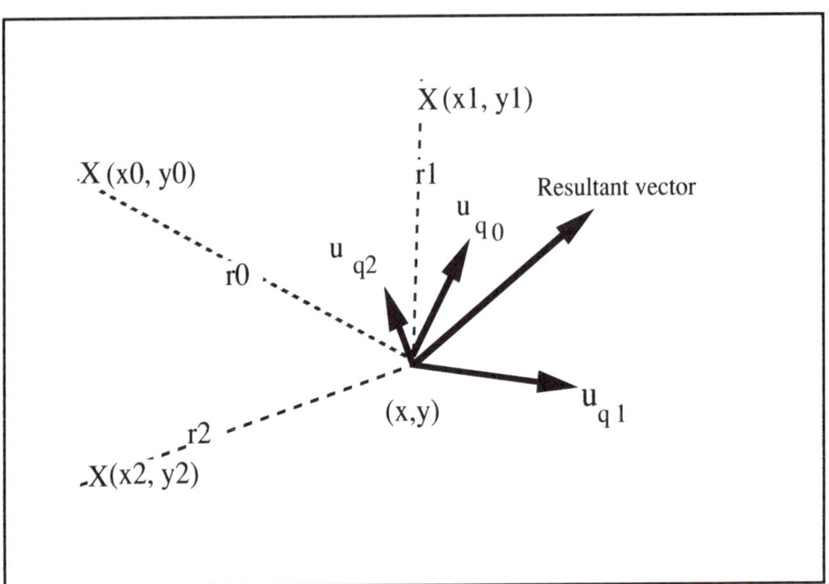

FIGURE 3.36. (a) This figure shows how a flow field can be created from its singularities by using vector summation

a proof, the reader is referred to [68, pp180-191].

$$\mathbf{v} = \nabla\phi + \nabla \times \mathbf{E}, \; where$$

$$\phi = -\frac{1}{4\pi}\int \frac{\Theta d\tau}{r}, \quad \mathbf{E} = \frac{1}{4\pi}\int \frac{\Omega d\tau}{r}$$

$$\nabla^2\phi = \Theta \tag{3.40}$$

This viewpoint is interesting, because it allows one to think of a vector field as being created through the placement of singularities with given intensities in space. In order to understand the relation between a flow field and its associated singularities, consider three vorticity singularities distributed in the xy plane, as shown in figure 3.36.

Let the three singularities be located at (x_0, y_0), (x_1, y_1), (x_2, y_2). We now wish to compute the velocity at point (x, y) that is induced by the three singularities. The velocity is given by the following equation, which involves vector summation

$$\mathbf{V}_{(x,y)} = \sum_{i=0}^{i=N} \mathbf{u}_{\theta_i} \tag{3.41}$$

where N is the number of singularities, and $\mathbf{u}_r$ and $\mathbf{u}_\theta$ are the radial and tangential components of the induced velocity due to a given singularity, defined by

$$\mathbf{u}_r = 0; \quad \mathbf{u}_{\theta_i} = \Gamma_i/2\pi r_i \tag{3.42}$$

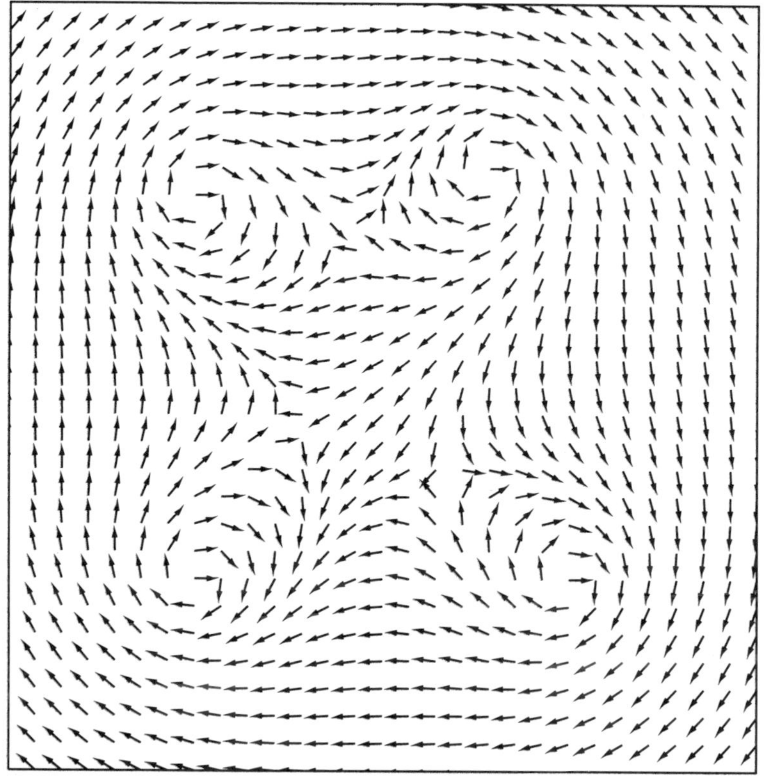

FIGURE 3.36. (b): The final flow field due to four singularities. The flow at each point is represented by arrows.

Here r_i is the distance from the singularity at (x_i, y_i) to the point (x, y). The sense of u_{θ_i} is taken to be perpendicular to the line joining (x_i, y_i) to (x, y), as shown in figure 3.36. Γ_i is the strength of the ith singularity. Hence, one can create a flow field by simply distributing singularities with given strengths across the (x, y) plane and summing vectorially the contributions due to each singularity at different points in the plane.

The reverse is also true – given a flow field, one can obtain the locations and strengths of each of the singularities that gave rise to the flow field. Once the singularities have been isolated, one can go back and construct the original flow field using equation 3.41.

3.8   Discussion

From the results on real images that we have presented so far, the power of using phase portraits to describe oriented patterns should become apparent. This should become clear when complex images such as that in figure 3.21 are considered. Even humans have a difficult time in deriving a qualitative description for what is happening in the texture. From this point of view, the algorithm does well in identifying the salient features of the texture that it is presented with. Based on these results, it appears that phase portraits are the appropriate models to use while describing oriented textures.

Another aspect is that the symbolic description that is ultimately derived is a rich description of the original texture. If we are only interested in a purely qualitative description, then it is easily available. If we want to go back from the qualitative description to the original texture, then the model makes available parameters (the **A** and **b** matrices) that allow a quantitative reconstruction of the phase portrait. Such a reconstruction closely resembles the original textures near central points. This constitutes a vast reduction in the amount of information that one needs to store about a given texture, and is a viable data reduction tool.

As pointed in section 3.2, orientation fields are very similar to optical flow fields. Hence, the method described in this chapter can directly be applied to optical flow fields in order to provide a segmentation as well as a symbolic description. In fact, the least squares problem becomes simpler if the input to our method is a vector field (see sections 3.2.4 and 3.4.1). This is because one does not have to deal with the ambiguity that is inherently present in orientation fields. In general, our method can be used to describe *any* kind of vector field, be it optical flow, fluid flow, or otherwise, in addition to orientation fields.

Our approach is currently computationally expensive on a serial machine. The non-linear least squares method takes approximately 15 seconds for a 11x11 window on an Apollo DN-3000 machine. For a 64x64 image, such a computation swept over the entire image would take several hours. However, this computation is parallel, as we are solving the same kind of least squares problem at different locations in the image. On a parallel machine like the Connection Machine, the computation time may be reduced to seconds.

Another point is that one can store the results of computing $\hat{\mathbf{A}}$ and $\hat{b}$ at every point in the image. These quantities represent an abstraction of the information present in the oriented texture, and all further computations are based solely on these abstracted values. Once this is done, one can experiment with various kinds of segmentation and reconstruction strategies, which take very little time. This is the approach that we used in our experimentation.

There are some limitations of this algorithm, which we would like to improve on in the near future. These limitations have already been pointed out in various sections, and are summarized here. First, the algorithm uses a linear approximation while classifying phase portraits. This is a necessary first step in order to gauge the usefulness of this algorithm. Once the validity of the algorithm is established, higher order approximations can be employed, such as a quadratic approximation. An arbitrary oriented texture is non-linear, and the only hope of analyzing it is to employ certain well-known approximations.

Second, the algorithm uses a fixed, predetermined window-size to estimate parameters of the phase portrait models. However the behaviour of the algorithm is not critically dependent on the choice of window-size. As mentioned, factors affecting the window-size are the validity of the linear approximation and the computational burden. Furthermore, instead of using a fixed windowsize, it would be nice to use a method similar to region growing, where one could start out with seed regions corresponding to fixed points of the different phase portraits. These seed regions could be grown until the phase portrait ceases to be linear. This idea is similar to the work done in [14].

Third, there are potential problems concerning robustness of the algorithm. Since we assume the leading term in $\mathbf{A}$ is either a 1 or 0 (Section 3.4.3.3), it is possible for the other entries to have extremely large values. This can be avoided by forcing the second, third or fourth entry to be a 1 or 0. What this implies is that different representations for $\mathbf{A}$ need to be used to generate the best solution. However, we did not carry this out in the interest of computation time. As demonstrated by the experimental results, this does not appear to be a critical factor in the performance of the algorithm.

We briefly discuss the expected effects of relaxing some of the simplifying assumptions in this paper. One of the assumptions is that there is a single dominant direction at each point in the texture. This assumption is used by the algorithm to derive the orientation field, and also by the algorithm to fit phase portraits. If this assumption is relaxed, it means that one allows multiple orientations to exist at a point, as in a cross-hatched pattern (e.g. wire netting). In such a case, the algorithm for orientation estimation will not work at those points where multiple orientations exist, such as the points of intersection of the cross hatches. These are actually singular points. If the singularity is isolated, then as one moves away from the singular point, there is a unique orientation for each flow vector. This will hold until another singular point is encountered. Since the algorithm to fit phase portraits relies on these orientation estimates, it follows that the success of the algorithm depends on the spacing between singularities. If the scale used to perform the fit is larger than the spacing between singularities, then an error is introduced.

This observation also holds for non-linear phase portraits that can be created by placing flow singularities next to each other, such as attractors and repellors [50].

Let us now consider flow discontinuities. An example of this is a herringbone pattern. In such a case, there are nonlinearities in the neighborhood of the discontinuity, and the method for fitting linear phase portraits will fail. This information could in fact signal the location of discontinuities, and could be a topic for further research.

3.8.1 EXTENSIONS TO THREE DIMENSIONS

Though the algorithms presented have been developed for the 2D case, they can easily be extended to the 3D case. In fact, it is the 3D analysis of velocity fields that is commonplace in fluid mechanics (see Ottino [94] for instance). In order to do this, one would have to first estimate the orientation field over a volume. Then the model $\dot{\mathbf{x}}(t) = \mathbf{A}\mathbf{x}$ could be fitted, where $\mathbf{A}$ is now a 3x3 matrix, and $\mathbf{x}$ a three dimensional vector.

3.9 Conclusion

In this chapter we presented a novel method to describe oriented textures. The scheme for analysis first computes an orientation field for flow-like textures. We reviewed a model for the description of flow-like textures, which is based on the qualitative theory of differential equations. The most important implication of this theory is that in a linear 2D case, there is only a finite number of qualitatively different phase portraits possible. We have used the mathematical terminology for these phase portraits as the set of symbols. This takes care of one part of the signal to symbol transformation problem.

We have also proposed computational techniques to determine how the transformation must proceed from the image to this set of symbols. The method consists of first computing the orientation field for the given texture. Consequently, a non-linear least squares technique is used to determine which phase portrait each portion of the texture resembles.

We have presented results of applying this technique to several real texture images. Specifically, we have segmented the given texture, derived its symbolic representation, and performed a quantitative reconstruction of the salient features of the original texture based on the symbolic descriptors. These results show that it is possible for a machine to automatically synthesize descriptions of complex oriented patterns. We have demonstrated the usefulness of this scheme by applying it to complex flow visualization pictures and defects in lumber processing.

4

Analyzing strongly ordered textures

4.1 Introduction

The central theme of this book is to provide symbols that will serve as a descriptive vocabulary for textures. The emphasis is on borrowing existing terminology from other disciplines, and then defining computational strategies to automatically extract these terms.

In this chapter we examine *strongly ordered textures*, also known as repetitive textures. These textures can be modeled by a primitive element that is replicated according to certain placement rules. In keeping with the spirit of this book, this chapter provides a vocabulary for primitives as well as placement rules.

Very few researchers have attempted to give an explicit symbolic description for repetitive textures. One such attempt is by Vilnrotter *et al* [120]. Later in this chapter, we will present a precise technique for providing symbolic descriptions. These symbolic descriptions are based on the equivalence classes of the frieze groups and the wallpaper groups.

The aim of this chapter is to take a critical look at the research done in the structural analysis of textures, with an eye for ultimately deriving accurate symbolic descriptors for strongly ordered textures. Thus, we provide directions for future research in ordered textures based on existing research.

A natural way of understanding strongly ordered textures is to examine the primitive elements, and the placement of these elements. Hence this chapter is organized as follows. Section 4.2 discusses the extraction of primitive elements and section 4.3 discusses the derivation of placement rules. Section 4.4 reviews existing models for strongly ordered textures. Sections 4.5 and 4.6 provide symbolic descriptions for both primitives and their placements which are borrowed from other disciplines, but are new to computer vision.

4.2 Extraction of primitives

The first step in the analysis of strongly ordered textures consists of the extraction of features, leading to the identification of primitive elements. Two broad classes of features are edge-based features and region-based features.

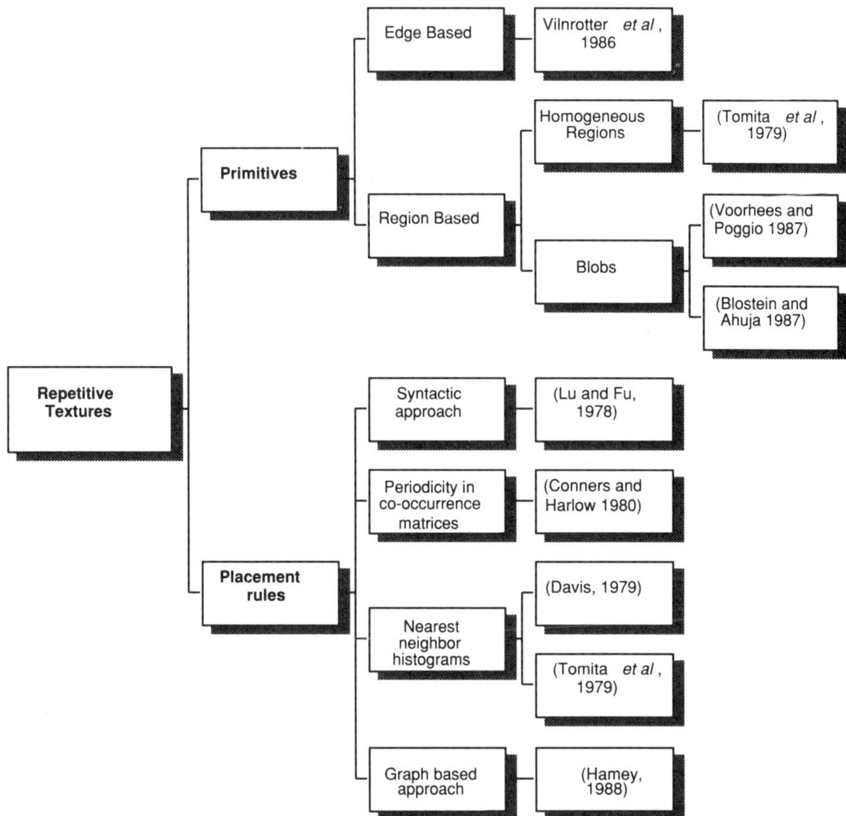

FIGURE 4.1. A summary of approaches to the structural analysis of textures

Edge-based features have been used by Vilnrotter *et al* [120]. Region-based features, also called *blobs*, are the more commonly used features [43, 122]. Figure 4.1 provides a summary of various approaches towards the structural analysis of textures.

4.2.1 EDGE BASED FEATURES

Vilnrotter *et al* [120] use edge based features in order to recover a symbolic structural description of an ordered texture. They propose the use of an *edge repetition array*, which is essentially similar to generalized co-occurrence matrices [27] for edges.

The edge repetition array is defined along six different directions in the image. It is basically a histogram of frequency of edge pairs of similar orientation versus the separation of such edge pairs. For a repetitive texture, these histograms will exhibit periodicity in the form of peaks at regular intervals. The major effort in their algorithm is directed towards describing and classifying these peaks, and subsequently extracting a simple symbolic description. An example of the symbolic description generated for a raffia texture along the vertical direction is as follows:

```
Vertical direction
    Strong evidence of periodicity (spacing 8.0)
    Strong evidence of element size (5.0)
    Moderate support for element spacing (8.0)
    Ratio of size to period: 0.63
```

Based on the analysis of peaks in the edge repetition array, it is possible to determine which edges are repetitive. The primitives for the texture are then extracted by filling in the region between repetitive edges along predetermined scan directions. Since the primitives are directional, they have to be merged to render a composite primitive. The algorithm seems to work well with natural textures.

Some of the shortcomings with the work of Vilnrotter *et al* [120] are that their approach involves quantization of *both* edge directions and the directions for the edge repetition arrays into a small set of directions. The effects of this quantization remain to be fully explored. In fact, the algorithm sometimes extracts unusual shapes for the primitive elements, as acknowledged by the authors [120]. This appears to be due to the quantization. Another problem with the algorithm that stems from quantization is that it will not be able to handle systematic distortions of an ordered texture, such as due to perspective projection. Finally, reconstruction of a given texture does not always agree with the original texture, as shown by the authors in [120].

4.2.2   Region based features

Earlier methods such as the one proposed by Tomita, Shirai and Tsuji [115, 116] assumed the texture element to be a region of homogeneous gray level. Simple region growing techniques were used to isolate these homogeneous regions. A simpler method is used by Leu and Wee [72], where the original textured image is thresholded to yield a binary image.

More recently, the favored initial stage of processing is filtering with the Laplacian of gaussian filter, $\nabla^2 G$. However, the interpretation of the response to $\nabla^2 G$ varies as follows.

4.2.2.1   Thresholding the response to $\nabla^2 G$

This is the approach used by Voorhees and Poggio [122]. They regard blobs as duals of edges, i.e., as areas in the image which do not correspond to edges. Since edges correspond to the zero crossings of $I \star \nabla^2 G$, blobs can be regarded as non-zero portions of the response to $\nabla^2 G$. In order to remove connections between blobs due to noise, a small threshold is used to isolate non-zero regions. Voorhees and Poggio present a technique to estimate the amount of noise in the image, which allows automatic setting of this threshold.

In order to avoid the problem of blobs being connected by narrow necks, they use a multiple thresholding technique, such that smaller blobs are extracted at higher thresholds. These smaller blobs are grown back over the original regions in order to segment them. Geometrical attributes such as length, width, and orientation are computed for each blob. Thus, no *a priori* assumptions about the shape of the blob are made.

The features of this algorithm that distinguish it from other blob detection techniques are that the processing is done at a single scale with multiple thresholds, and that no assumptions are made about the shape of the blob.

4.2.2.2   Solving for the response to a disk at multiple scales

This is the approach used by Blostein and Ahuja [16], and Hamey [43]. Here, the assumption made is that a blob is a circular region of constant intensity. Hence, the approach involves analyzing the response of $\nabla^2 G$ to a circular intensity disk, and conversely recovering the diameter and contrast of the disk from the response.

The algorithm is based on calculations of the $\nabla^2 G$ and $\frac{\partial}{\partial \sigma}\nabla^2 G$ responses of a disk image. A closed form solution for the response at the center of the disk is available, which enables the recovery of the diameter D and contrast C of the disk as follows:

$$
\begin{aligned}
D &= 2\sigma\sqrt{\sigma(\frac{\partial}{\partial\sigma}\nabla^2 G \star I)/(\nabla^2 G \star I) + 2} \\
C &= \frac{2\sigma^2}{\pi D^2}e^{D^2/8\sigma^2}(\nabla^2 G \star I)
\end{aligned}
\tag{4.1}
$$

where the convolutions are evaluated at the center of the disk. The above equation is applied at the local extrema of $\nabla^2 G \star I$. The justification is that at the proper scale, a local extremum corresponds to the location of a disk center.

Since there could be disks of different diameter in a given image, the algorithm employs a bank of $\nabla^2 G$ filters at different scales. Equation 4.1 is regarded as meaningful at a given scale σ only if $\mid D - 2\sqrt{2}\sigma \mid < 2$ *pixels*. This provides a criterion for combining the output from several scales.

Hamey [43] extends this approach to selectively discard false responses to ridges and edges in the image. However, this involves the *a priori* setting of a threshold which could be done automatically by using the algorithm of Voorhees and Poggio [122].

The distinguishing features of the response-to-a-disk technique are that it assumes the shape of blobs to be circular, and that the analysis is done at multiple scales of filtering.

4.3   Extracting structure from primitives

As Hamey [43] pointed out, understanding a repetitive pattern is a cyclic problem: one needs to determine the primitive to derive the placement rules, but the primitives are in turn decided by the nature of repetition in the pattern. Most approaches implicitly extract the primitives first, but Hamey makes this process explicit through the introduction of a dominant feature assumption. That is, the most dominant feature is first extracted, and then the repetitive structure of these dominant features is derived.

In this section we summarize approaches to extracting structure from the primitives that are found in the first stage of processing.

4.3.1   SYNTACTIC APPROACHES

Lu and Fu [77] adopt a syntactic approach to analysis of structured textures. The primitive element is not defined explicitly, but is actually a single pixel. Structure is extracted by determining whether a given texture belongs to the language generated by some *a priori* grammar. This grammar then defines the given texture.

The original image is subdivided into windows of prescribed size. For each window a tree of pixel values is constructed, and is compared with a tree generated by a predefined grammar, which makes the approach essentially brute force. This method was applied for binary images only, and due to combinatorial explosion, would be impractical for gray level images. As the authors pointed out, the grammars are very sensitive to noise, and this casts serious doubts on the usefulness of the algorithm. Furthermore, constructing a grammar to describe a real texture is very difficult. Finally, the algorithm is critically dependent on the choice of the window size, a problem that has not been addressed by the authors.

4.3.2   NEAREST NEIGHBOR HISTOGRAMS

Davis [28] presents a method to compute the spatial structure of texture elements. The method deals only with texture patterns consisting of dots. The method involves a heuristic procedure to measure repetitiveness in the histogram of the distance between a point and its k-nearest neighbors. The problem of finding the structuring element has not been addressed. Further, only rectangular and square grids are considered.

Tomita, Shirai and Tsuji [115, 116] classify primitives based on properties such as brightness, area, elongatedness and curvature. The relative positions between the centroids of the elements is used to derive a detailed placement description. This is done by computing the histogram of the relative positions between every pair of elements. As pointed out by the authors, the algorithm has the limitations that the segmentation scheme is weak and the method for deriving placement rules is too simple.

Leu and Wee [72] use a variant of the nearest neighbor histogram method by employing clustering analysis to recover the two vectors that describe repetition.

4.3.3 USING CO-OCCURRENCE MATRICES

Most of the research done on texture analysis has focussed on using co-occurrence matrices. Conners and Harlow [24] describe a method for the structural analysis of textures using features defined on co-occurrence matrices. They define an inertia measure on the co-occurrence matrix, and illustrate the use of this measure to detect periodicity in the texture. Similar results can be obtained by defining measures directly on the two dimensional autocorrelation function, as shown in [104].

There is a close similarity between detecting periodicity in measures defined on the co-occurrence matrix and analyzing nearest neighbor histograms, as in the earlier section. The heart of such an analysis is the characterization of periodicity and peakiness in histograms.

4.3.4 GRAPH BASED APPROACHES

Hamey [43] provides an elegant and robust formulation for the problem of extracting structural descriptions from primitive elements. Previous approaches to this problem were intuitive without rigorous justification. Hamey provides a formalization of these intuitive ideas, and the mathematical significance of using the nearest neighbor approach used previously is demonstrated. Earlier approaches relied on histogramming techniques [116, 120] and statistical methods for the extraction of a structural description. Hamey uses explicit knowledge of the structuring process in order to derive his algorithm.

The basic idea in Hamey's work is to construct a neighborhood graph for features and extract repetitive links from this graph based on constraints from lattice theory and feature prominence. It must be pointed out that the results that Hamey uses from lattice theory are derived differently and better known in the areas of symmetry [42, 86, 126] and crystallography [132].

The algorithm begins by grouping features that are likely to be related. This is done by assigning weights to the links in a graph that has features as nodes. The assignment of weights depends on the dominance of the features, and the proximity of interfering features. In the next stage, only strong links are retained to determine local repetitiveness. This is done by considering three types of local repetitiveness, viz. cross, T and L shaped repetition. An evaluation function based on the perpendicularity of fundamental frequency vectors and agreement with data is used to retain the strongest repetitions. Finally a relaxation algorithm is applied to resolve incompatibilities.

The algorithm appears robust and performs much better than previous approaches. Hamey's work concentrates mainly on the extraction of local repetition. The problem of detailed description of primitives during feature detection still remains to be addressed. Due to blob detection, the algorithm does not compute shape parameters. Hence certain kinds of repetitive structures cannot be handled by the algorithm as pointed out by the author [43, p 70]. In fact, in section 4.6 we will show a very interesting direction that Hamey's work can be continued in if the blob features are analyzed in more detail.

4.4   Models for strongly ordered textures

There are only a few explicit models for strongly ordered textures in the computer vision literature. In this section we briefly discuss these models.

Zucker [134] describes a model for repetitive textures based on the notion of primitives, ideal textures and observable textures. The primitives are restricted to regular polygons. Ideal textures are generated by the concatenation of primitives in such a way that each primitive is surrounded identically. This makes Zucker's class of ideal textures very restrictive. Zucker formulates the tiling problem as a graph theoretic problem. The novel feature of Zucker's work is to propose a model for distorting an ideal texture into an observable texture, eg. such as perspective distortion. However, no implementation results are presented in this paper.

Ahuja and Rosenfeld [1] propose a class of image models, called mosaic models, which are based on random geometric processes. Two types of models are considered: cell structure models and coverage or 'bombing' models. Cell structure models involve the tessellation of a plane into bounded convex polygons. Different techniques of tessellating the plane give rise to different classes of cell structure models (e.g. checkerboard model, triangular model). A coverage mosaic is obtained by a random arrangement of a set of geometric figures in the plane. The random arrangement is modeled by a Poisson point process which drops the centers of the figures onto the plane, and the figures considered are line segments, ellipses, circles, rectangles and squares.

One can compute the expected values of properties such as edge density, perimeter, and number of cells in an ideal image generated from the model. The same properties can also be estimated from an actual image, and this is the basis of a scheme to fit the appropriate mosaic model to a given texture.

Hamey [43] analyzes only the aspect of repetition in textures. He proposes three models of repetition as follows. The *grid model* of repetition is defined by two basis vectors with a placement distribution error. This error causes grid points to move around by a small amount. The *consistent relative placement model* is similar, except that the distribution error is allowed to be cumulative. This can result in a phase drift of the repetition. The *local repetition model* uses only locally defined basis vectors, which are allowed to vary across the image.

4.4.1 Directions for further research

Computer analysis of strongly ordered textures is able to produce a description involving simple placement rules. Interestingly, a concrete classification of different textures is available if further effort is invested in a direction specified shortly. The theory of transformation geometry and symmetry provide the appropriate models for ordered textures. This is a topic which has been well investigated in mathematics, and has potential for applications in computer vision. (Though early approaches to texture, such as Zucker's [134] mention aspects of tiling theory and symmetry, no implementations were proposed.)

The idea is to derive descriptions for textures such that they fit into equivalence classes. In this way, even though there are an infinite number of natural textures, they are all classified into a finite number of easily understood equivalence classes. This approach is essentially similar to the one used for classifying different kinds of flow-like textures based on their phase portraits.

4.5 Symbolic descriptions: models from petrography

In this section we present a model for the symbolic description of primitives and placement rules from the area of petrography [110, 78]. This scheme is useful for strongly ordered textures that do not exhibit symmetry. Such textures are found in metallography [123] and petrography [110, 78]. Section 4.6 discusses the appropriate model when the texture exhibits symmetry.

4.5.1 Description of primitives

Most algorithms do not place much emphasis on the description of primitives. Admittedly this is a difficult task, as primitives can come in a variety of shapes and sizes. Some approaches such as Tomita *et al* [116] characterize primitives by geometrical properties such as area, elongation and orientation. However, these descriptions are not symbolic. Other approaches to texture avoid the issue of the description of primitives by treating them as blobs, such as Hamey's work [43]. Of course, such an approach is initially justifiable if one wants to extract placement rules. However, once the repetitive blobs are identified, it remains to provide good symbolic descriptors for the primitives.

Fortunately, in the disciplines of metallography and petrology, this is

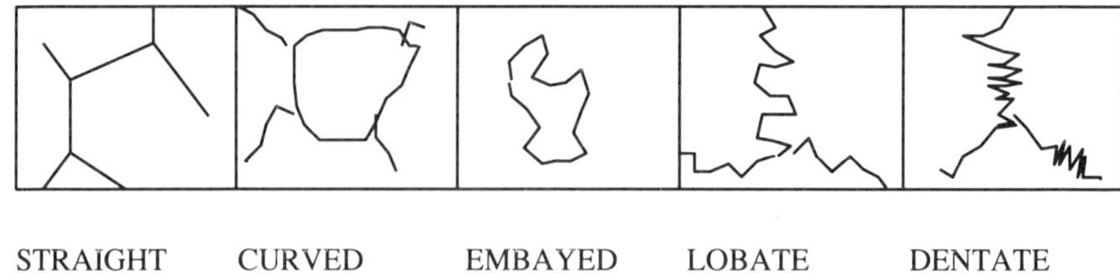

Symbolic descriptors associated with grain boundary shapes

FIGURE 4.2. Symbolic descriptors for primitives such as grains are available in the area of petrography

precisely one of the problems that has received considerable attention. Researchers in these areas have already developed an accurate characterization and description of primitives called *grains* or *crystals* [110, 78]. It will not be possible to provide an exhaustive survey of such descriptors, but a brief discussion will be included to provide a flavor of the kinds of descriptors available.

4.5.1.1   Terminology from petrology

Symbolic descriptors are used in the area of petrology for different primitives in the form of grain shapes. For instance, Spry [110] uses the following terms to describe grain boundary shapes: straight, curved, embayed, scalloped, lobate, dentate, unilateral and bilateral. Figure 4.2 illustrates a few of these shapes.

The reason for showing this particular example is to emphasize the fact that one need not start from scratch in the search for symbolic descriptors. Several descriptors are available in other disciplines, which are very relevant to solving the signal-to-symbol transformation problem. This problem is one of the central issues in computer vision.

4.5.2   DESCRIPTION OF PLACEMENT OF PRIMITIVES

A precise vocabulary is available for aggregations of primitives. The importance of aggregate distributions arises because it may not always be possible to find vectors that define repetition in a texture. Crystals in a rock can be *equigranular* (all crystals of approximately same size) or *inequigranular* (crystals differ substantially in size). To get a flavor for the different possibilities, we present a description used by MacKenzie [78] for the inequigranular case. This category includes seven kinds of texture:

1. *Seriate* - crystals of the principal minerals show a continuous range of sizes

2. *Porphyritic* - relatively large crystals are surrounded by finer-grained crystals of the groundmass

3. *Glomeroporphyritic* - there are crystal clots or clusters

4. *Poikilitic* - Relatively large crystals of one mineral enclose numerous smaller crystals of one or more other minerals

5. *Ophitic* - a variant of poikilitic texture in which randomly arranged crystals are elongate and are wholly or partly surrounded by larger crystals

6. *Interstitial textures* are classified based on the material occupying the spaces between crystals

7. *Oriented, aligned and directed textures* Several classes of this textural type exist such as trachytic, trachtoid, comb texture, orbicular texture. Rather than there being a single universal alignment direction, there can be several domains, each having its own preferred direction of alignment.

Note that these are precise symbolic descriptors for different kinds of aggregations of primitive elements. Though these symbolic descriptors have been devised for petrology, a similar approach could be used to describe textures from other domains also.

4.6   Frieze groups and wallpaper groups

In this section we describe a model for symmetric ordered textures, based on the frieze groups and wallpaper groups. This model involves analyzing the symmetry of the structuring element in relation to the placement rules. Though the model is known in the area of transformation geometry and symmetry, its introduction to the computer vision literature is new.

Early approaches to structural analysis of textures [134] introduced the notion of ideal textures. Real textures were then considered to be distortions of these ideal textures, where the amount of distortion could vary. In the spectrum of possible distortions, researchers have shied away from distortion-free ideal textures on the grounds that *real* textures are heavily distorted [43]. Thus, there seems to be an implicit notion in the minds of researchers that real textures are necessarily distorted or random. This is only *partly* true, because there are a large number of naturally occurring textures in the real world which are ideal or near ideal. In areas such as analysis of diffraction patterns, and scanning electron microscopy for materials analysis, there are several textures that exhibit an ideal behaviour [88, 132].

It is interesting to note that natural textures are used to motivate the study of geometrical symmetry [126, 75]. A convincing argument is provided in [75, p. 2], as follows:

> The physical process of growth, be it of crystals or of living matter, favors the production of symmetrical forms; and indeed the world is pervaded by shapes having a greater or lesser degree of symmetry, from the leaf to the human form ....

> Two dimensional plane groups describe the decorative patterns applied to surfaces, as seen in wallpapers, tessellated pavements, mosaics and tiled arabesques, and the repetitive patterns of many carpets and rugs.

Thus, for a complete understanding of natural textures, one must analyze *both* the ideal case and the distorted case. The analysis of ideal textures has been prematurely abandoned by researchers in computer vision.

Based on our viewpoint of designing a vocabulary for texture, we would like to emphasize the importance of examining symmetrical textures closely. The main reason for this is that there is an elegant mathematical theory that provides a clean classification for all symmetrical textures. A classic result in this domain is that there are exactly seventeen classes of wallpaper patterns.

Based on the literature survey carried out in this chapter, it is clear that considerable research remains to be done on the analysis of strongly ordered textures. It appears that the methods proposed so far work reasonably well in extracting primitive elements [16, 122] and deducing placement rules [43]. What remains to be done is to derive appropriate symbolic descriptors for the texture. The purpose of this section is to examine the discipline of transformation geometry in order to understand the description scheme used in this area.

It will be virtually impossible to give a detailed introduction to the topic of transformation geometry, but only the ideas relevant to computer vision will be presented here. For proofs and exposition of the theory, the reader is referred to [64, 86, 75]

4.6.1 BACKGROUND

Though the topic of tiling theory [42] was briefly mentioned by Conners and Harlow [24], it was used only to define the unit patterns of periodic textures. The full power of the subject of transformation geometry and symmetry was not brought to bear on the structural analysis of textures. We approach this subject from a different viewpoint: the description of structural textures must *produce* a result which falls into one of the mathematically defined equivalence classes for the tiling patterns. Once such a characterization of the given texture is available, it forms a complete description of the given

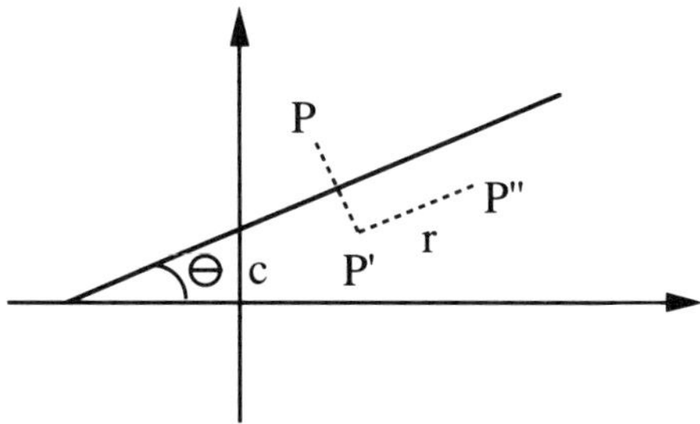

A glide reflection: Point P is first reflected in the line
y = xtan θ + c to produce P', which is then translated
parallel to the line by a distance r to produce P"

FIGURE 4.3. Illustrating the glide reflection, which is one of the isometric transformations

texture. Such a direction has not been mentioned in the literature, and it
appears feasible, given the encouraging results obtained in [43].

4.6.2   PRELIMINARY DEFINITIONS

A set of elements together with an operation satisfying the properties of
closure, associativity, existence of an identity element, and existence of an
inverse for every element is called a *group*. A *transformation* is a one-to-one
mapping of a set into itself. Note that the set of all transformations forms
a group under the operation of multiplication of transformations. We shall
be concerned with transformations on the plane, ie a mapping of the plane
onto itself.

We now restrict our analysis to *affine transformations*. An affine transformation from the set of all ordered pairs (x, y) of real numbers to the set
of ordered pairs (x', y') of real numbers is given by the following equation

$$x' = ax + by + h, \quad y' = cx + dy + k, \quad ad - bc \neq 0 \qquad (4.2)$$

The importance of the group of affine transformation arises from the fact
that affine transformations are the *only* one-to-one mappings of the plane
onto itself which always map straight lines onto straight lines [64, p. 60].

From the group of all affine transformations, we restrict our attention to
those transformations that preserve distances, known as *isometric transformations* or *isometries*. The study of the group of isometric transformations
is based on the fixed points of the transformations. An important result is
that the only isometric transformations are rotations, translations, reflections and glide reflections. Figure 4.3 illustrates a glide reflection.

We now focus our attention on symmetry. A point O is called a *center
of symmetry* [64] for a given figure F if, whenever a point P belongs to

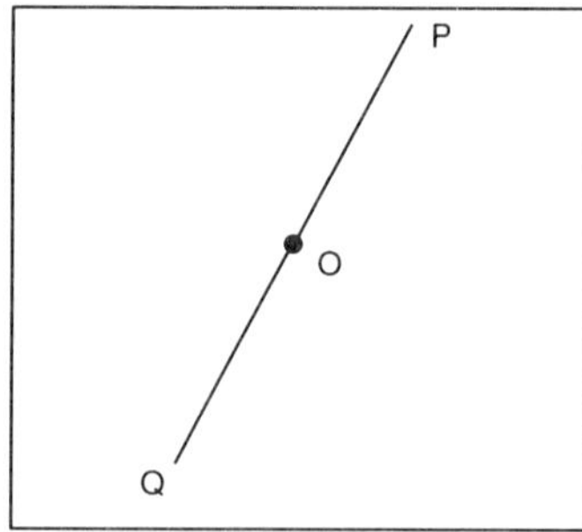

FIGURE 4.4. Defining the center of symmetry

figure F, point Q also belongs to the figure F, where point Q is obtained by producing PO and taking OQ = PO, as shown in figure 4.4. A figure is said to have a *central symmetry* or a *point symmetry* about point O, if O is a center of symmetry for the figure.

Let us turn to groups of symmetry. An isometry is said to be a *symmetry* of a figure if it brings the figure to coincidence with itself. The set of all possible symmetries of a figure forms a group (since if symmetries S_1 and S_2 bring a figure to coincidence with itself, then so does S_2S_1, and so on)

The classification of symmetries of all plane figures is based on the translational structure. Three cases arise:

1. When no translations are present, the groups of symmetries are called *point groups*. Thus a point group can contain only reflections and rotations.

   The group of symmetries for a regular polygon of n sides is called the *dihedral group* of order $2n$, and is denoted by D_n. The group D_n consists of n rotations and n reflections. The rotations alone form a subgroup of order n, called a *cyclic group*, and denoted by C_n.

   It can be shown [64] that the only possible point groups are given by C_n and D_n, for $n = 1, 2, 3, ...$

   There are two infinite sets of symmetry types for finite patterns in two dimensions, the cyclic types and dihedral types.

2. When only one direction of translation is present, the groups of symmetries are called *frieze groups*. A repeating pattern is built up by translations, which move every point in the same direction by the same distance. A *frieze pattern* consists of a motif (an element of the pattern) which is repeated at regular intervals along a line *ad infinitum*. Another way of looking at a frieze pattern is to see that it is left fixed by some "smallest translation."

3. When more than one direction of translation is present, the symmetry groups are called *wallpaper groups* and there are in all seventeen classes of wallpaper groups. In this case, the repeating pattern consists of a motif repeated at regular intervals in more than one direction.

4.6.3 FRIEZE GROUPS

We have seen that a frieze pattern consists of a motif which is repeated at regular intervals along a line. Other symmetries in addition to translation can also be present. Note that there is an infinite variety of frieze patterns possible, because the motif can be selected arbitrarily. In spite of this variation, an intriguing result is that *there are only seven possible types of ornamental frieze patterns.* This is obtained by discounting scale factors and the subject matter of the motif, and considering only the symmetries which leave the pattern invariant.

This is a valuable result from the point of view of computer vision, because it allows an elegant scheme to classify any given frieze pattern into one of the seven equivalence classes. Again, we stress the need for appropriate models in computer vision; for ornamental patterns, the frieze groups are the appropriate models. This is an important idea, because it is similar in spirit to classifying different phase portraits into canonical equivalence classes, as presented in the chapter on analyzing oriented textures.

The seven frieze groups are illustrated in figure 4.5.

4.6.4 WALLPAPER GROUPS

A *wallpaper pattern* is one consisting of a motif repeated at regular intervals in more than one direction. When a discrete group of isometries of the plane does not stabilize either a point or a line, the group is called a *wallpaper group*.

Let W be a wallpaper group, and α be a transformation in W. Then a smallest polygonal region t such that the plane is covered by $\{\alpha(t) \mid \alpha \in W\}$ is called a (polygonal) *base* for wallpaper group W [86]. The bases can be used to generate a wallpaper pattern having a given wallpaper group as its symmetry group. This is the connection between wallpaper groups and wallpaper patterns.

The key step towards classifying the wallpaper groups is to establish the *crystallographic restriction.* According to this restriction, the only possible rotations have order 2, 3, 4 or 6. The consequence of this is that a wallpaper group has one of ten different point groups, C_1, C_2, C_2, C_4, C_6, D_1, D_2, D_3, D_4 or D_6. The analysis of each of these possibilities completes the classification of wallpaper groups into seventeen classes. Figure 4.6 provides illustrations of the different wallpaper groups.

The importance of the wallpaper groups is that they can serve as a set

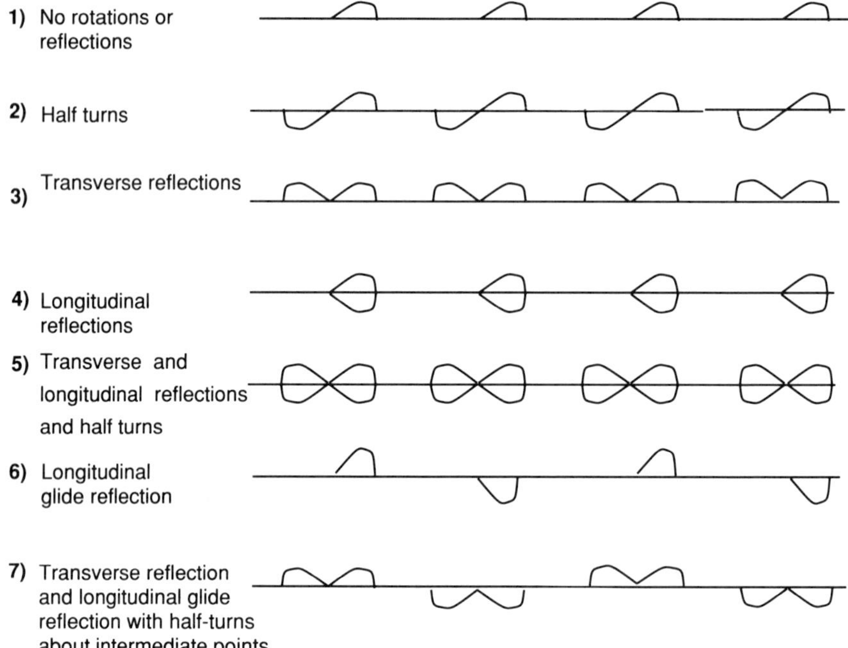

FIGURE 4.5. Illustrating the seven frieze groups

1) Only translations
(no odd isometries)

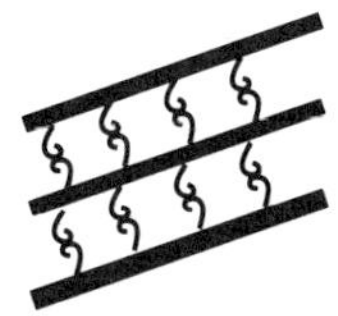

2) Translations and half-turns

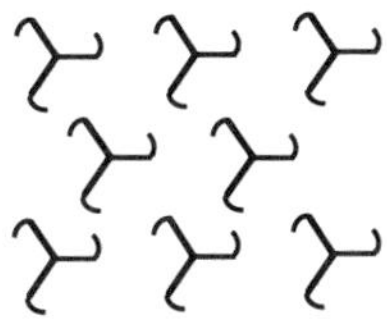

3) Translations and rotations
through 120 or 240 degrees
(no lines of symmetry)

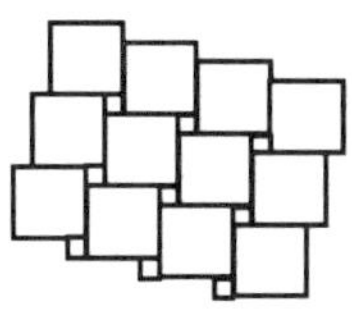

4) Translations and rotations
through 90, 180 or 270 degrees
(no lines of symmetry)

5) Translations and rotations
through 60, 120, 240, 300, 360
degrees (no lines of symmetry)

6) No n-centers: All axes of
glide reflection
 are axes of symmetry

7) Translation and glide
reflection
(no line of reflection but glide
 symmetry)

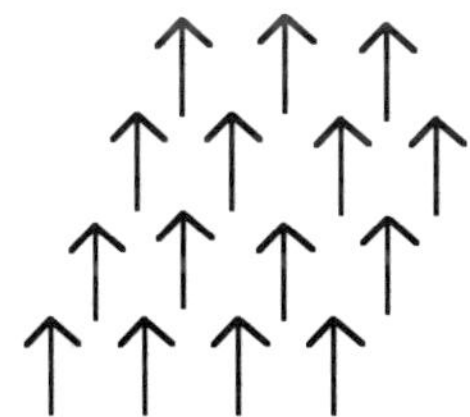

8) Some axes of glide
reflection
are not lines of symmetry

FIGURE 4.6. Illustrating the different wallpaper groups

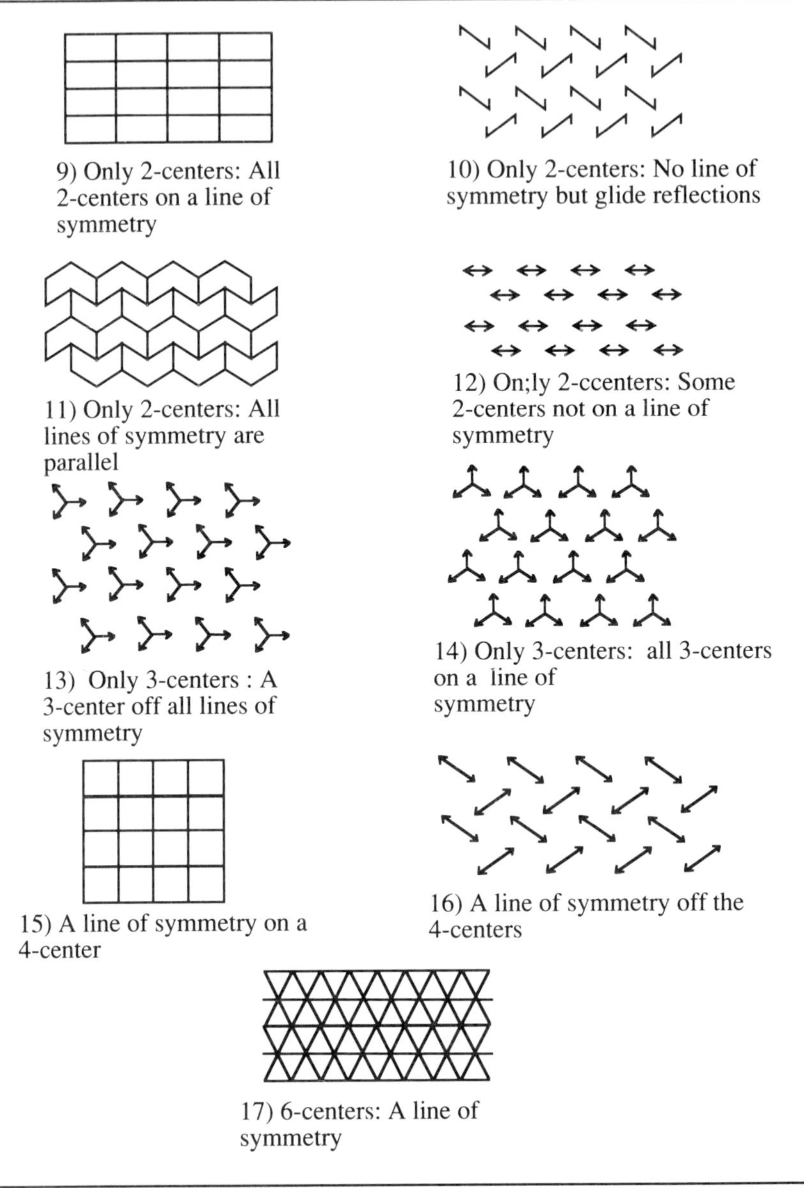

FIGURE 4.6. (contd) Illustrating the different wallpaper groups

of equivalance classes into which *any* repetitive texture can be classified. Thus, the description of a repetitive texture is mathematically complete as soon as the wallpaper group it belongs to is determined.

The reader is referred to [42] for recent advances in the related area of tiling theory.

4.7   Implications for computer vision

Having presented a classification scheme for structural textures, the issue that remains to be addressed is: how can an automatic classification of a given pattern into one of these classes be performed? Lockwood and Macmillan have suggested ways for doing this, in the chapter entitled 'Classifying and identifying plane patterns', in [75]. The basic idea is to identify a motif or structuring element, which determines the repetition of the texture. Then, symmetries of the structuring element are identified, which enable a classification of the texture into one of the seventeen wallpaper groups.

The problem thus becomes one of identifying the primitive element, the directions of replication, and finally the symmetries of the primitive element. As shown in sections 4.2 and 4.3, this is a problem that seems well studied in computer vision. The next step is to perform the analysis suggested by Lockwood and Macmillan [75] above. Brady and Asada [18] have described an algorithm to detect smoothed local symmetries in two dimensional objects. This method applies to the two dimensional contour of an object. More recent work in the detection of symmetry based on perceptual grouping can be found in [90]. Further analysis of the symmetry can be performed by examining the gray level variations within the object. The analysis of the symmetry of the primitive element, and its relation to the placement rules (as illustrated in figure 4.6) completes the description of a repetitive texture.

Such a system would be very useful in crystallography , where the automatic classification of symmetry types is still an unsolved problem. Researchers in the area of crystallographic analysis regard its automation through machine vision as significant. In fact, computer aided microscopy is becoming a new research area, which involves the application of image processing and computer vision techniques to aid the interpretation of images acquired through various types of microscopes. Especially in the domain of scanning electron microscopy applied to materials analysis, there are a large number of images that can be analyzed as strongly ordered textures (see [138] for instance).

4.8   Summary

In this chapter we first reviewed some of the important work done in the area of structural analysis of textures. The ultimate goal of such an analysis is to derive a symbolic description of a given strongly ordered texture. Researchers appear to have solved only part of the problem, viz. to identify the primitive element, and the repetitive placement rules.

What remains to be done is to derive proper symbolic descriptions from currently available results. This chapter has presented symbolic description schemes for the primitive element together with the placement rules. Symbolic descriptions for primitive elements are available in areas such as petrology, and pointers have been provided to the literature. Another set of symbolic descriptions for symmetrical textures is provided by the frieze groups and wallpaper groups. These groups have been illustrated, and a strategy for identifying members of the groups has been presented.

5

Disordered textures

Disordered textures are those that show neither repetitiveness nor orientation. There are several methods for analyzing disordered textures, based on features such as roughness, edgeness, entropy and contrast. In this chapter we briefly review some of these methods.

5.1 Statistical measures for disordered textures

Haralick [46] classifies statistical approaches to texture into eight categories: autocorrelation functions, optical transforms, digital transforms, textural edgeness, structural elements (in morphology), spatial gray tone cooccurrence probabilities, gray tone run lengths and autoregressive models. Since these approaches have been reviewed in the literature [46, 125], we will not discuss them in detail.

Instead, we will focus on some measures that have become popular in the literature. The most widely used of the statistical approaches is that of the co-occurrence matrices, or the spatial gray level dependence matrices. Conners and Harlow [24] compare the following texture algorithms: spatial gray level dependence matrix, gray level run length method, gray level difference method and the power spectral method. They conclude that the spatial gray level dependence matrix is the best in the sense that it retains the most texture-content information. Texture content information is defined as follows: a texture algorithm is said to experience a loss of important texture-content information if there exist two visually distinct textures which cannot be discriminated by the algorithm [24].

In the context of analyzing disordered textures, we briefly consider the use of entropy, as derived from the co-occurrence matrices.

5.1.1 COMPUTING THE ENTROPY AS A MEASURE FOR DISORDER

In order to define the entropy of an image, we first have to compute the co-occurrence matrices. Consider the second-order joint conditional probability density function, $f(i, j \mid d, \theta)$. For a given θ and d, $f(i, j \mid d, \theta)$ represents the probability of going from gray level i to gray level j, given that the intersample spacing is d and the direction of the intersample spacing is θ. For a given value of d and θ we can generate a matrix which represents the estimated second order joint conditional probability density function, which is known as the *gray level co-occurrence matrix* [45].

Let p_{ij} be the $(i,j)th$ entry in the cooccurrence matrix. The entropy is defined as

$$s = -\sum_i \sum_j p_{ij} log(p_{ij}) \tag{5.1}$$

In section 5.4.1, we present experimental results of using the above measure on some images.

5.2 Describing disordered textures by means of the fractal dimension

An alternate technique for characterizing a disordered texture is to measure its fractal dimension. This idea was suggested initially by Pentland [97], for itensity images. The following technique is also being used to analyze natural surface shapes [131].

A random function $I(x)$ is a fractal Brownian function if for all x and $\triangle x$

$$Pr(\frac{I(x + \triangle x) - I(x)}{\parallel \triangle x \parallel^H} < y) = F(y) \tag{5.2}$$

where Pr is the probability and $F(y)$ is a cumulative distribution function [81].

$I(x)$ describes a function whose fractal dimension is

$$D = T + 1 - H \tag{5.3}$$

where T is the topological dimension of $I(x)$.

5.2.1 ADVANTAGES

The fractal dimension of the texture constitutes a useful measure of roughness. This is because studies have shown [97] the fractal dimension to correlate very well with a human's assessment of surface roughness. This aspect is important to consider when one is designing practical vision systems to be used by novice users [104]. As we show in section 5.4.1, measures defined on the co-occurrence matrix do not share this property.

The fractal dimension defines how jagged or crumpled a surface is, $D = 2$ corresponding to a flat plane and $D = 3$ corresponding to a highly spiked surface. Furthermore, the fractal dimension of a surface is invariant with respect to linear transformations of the data and to transformations of scale.

5.3   Computing the fractal dimension

Pentland [97] has shown that if a natural surface is fractal then the image intensity is also fractal. To estimate the fractal dimension of the surface that the image represents, one can interpret $I(x)$ as being the image intensity. This allows us to measure the fractal dimension using different techniques, of which we pick the following two methods for the sake of comparison.

5.3.1   USING THE EXPECTED VALUES OF INTENSITY DIFFERENCES

Rewrite equation 5.2 as

$$E(|\ \Delta I_{\Delta x}\ |)\ \|\ \Delta x\ \|^{-H} = E(|\ \Delta I_{\Delta x=1}\ |) \tag{5.4}$$

where $E(|\ \Delta I_{\Delta x}\ |)$ is the expected value of the absolute value of the intensity change over scale Δx. From equation 5.4 we obtain

$$H = [log(E(|\ \Delta I_{\Delta x}\ |)) - log(E(|\ \Delta I_{\Delta x=1}\ |))]/log(\Delta x) \tag{5.5}$$

One can perform linear regression to estimate H, which is related to the fractal dimension of the surface.

Results of using this technique are presented in section 5.4.2.

5.3.2   THE RETICULAR CELL COUNTING METHOD FOR COMPUTING THE FRACTAL DIMENSION

An alternate method for measuring the fractal dimension [36], called the 'reticular cell counting' approach was found to be quicker than the above method, (as shown in section 5.3.3) and also provided satisfactory results.

Consider the surface that the image intensity values represent. Consider a given scale λ, and form subcubes each of whose sides is equal to λ (see figure 5.1). Now count the number $N(\lambda)$ of subcubes that contain at least one sample of the intensity surface. By Mandelbrot's relation [80]

$$N(\lambda) = \lambda^{-D} \tag{5.6}$$

where D is the fractal dimension of the surface being measured.

This gives us an algorithm for measuring the fractal dimension of a surface. We choose several scales, and find $N(\lambda)$ for each scale. By fitting a line to the plot of $log(N)$ versus $log(\lambda)$ and measuring its slope, we obtain the estimate of the fractal dimension of the surface. Specifically, if we let $Y = mX + c$ denote the equation of the least squares fit line, where Y corresponds to $log(N(\lambda))$ and X corresponds to $log(\lambda)$, then the fractal dimension $D = -m$.

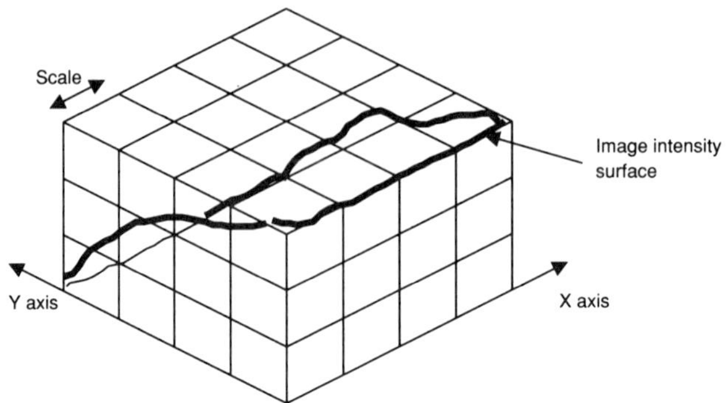

FIGURE 5.1. Illustrating the process of measuring the fractal dimension. The number of cubes that contain at least one image intensity point is calculated. The size of the cube is equal to the scale being considered.

Let Y_i denote the ith data point, $\hat{Y}_i$ denote the fitted value, and $\bar{Y}$ denote the mean of the fitted values. The correlation coefficient r is defined as follows [92]

$$r^2 = \Sigma(\hat{Y}_i - \bar{Y})^2/\Sigma(Y_i - \bar{Y})^2 \tag{5.7}$$

The correlation coefficient r is used as a confidence measure in the fractal dimension computed by the algorithm – a value of r close to 1 indicates that the fit is linear, and hence that the fractal dimension is a reliable estimate of the surface's roughness.

This method has been implemented, and the results are shown in section 5.4.3

5.3.3   A COMPARISON OF THE METHODS FOR MEASURING FRACTAL DIMENSION

1. Consider the method using expected values of intensity differences, in section 5.3.1. Assume a given scale i. Each point in the image will have $8i$ neighbors at a distance i from it. The value $E(|\ \Delta I_{\Delta x}\ |)$ is calculated, where $\Delta x = i$. Assuming the image is of size $N\mathrm{x}N$, this will require $8iN^2$ computations. Assuming that scales ranging from 1 to M are used in the estimation of the fractal dimension, the time complexity of this algorithm is

$$\sum_{i=1}^{M} 8iN^2 = O(M^2N^2) \tag{5.8}$$

2. For the reticular cell counting method in section 5.3.2, consider a given scale i. During the first pass, each point in the image is assigned

to a cube. This takes N^2 computations. Then we must determine the number of cubes containing at least one image point. Thus we must count over all the cubes in the worst case. Assuming there are G gray levels in the image, this requires $(G/i)(N/i)^2$ computations. Thus, for M scales, the total computation time is

$$\sum_{i=1}^{M}(N^2 + GN^2/i^3) \tag{5.9}$$

In order to evaluate this, consider Riemann's Zeta function, which by definition is

$$\zeta(p) = 1 + 1/2^p + 1/3^p + 1/4^p + ... = \sum_{k=1}^{\infty} k^{-n} \tag{5.10}$$

We are interested in $\zeta(3)$ which is $\simeq 1.202$. Thus, the time complexity in the worst case is

$$\sum_{i=1}^{M}(N^2 + GN^2/i^3) = O(MN^2 + 1.202GN^2) \tag{5.11}$$

Clearly, for large values of M and N, the second method is superior.

5.4   Experimental Results

We now present the results of using the algorithms in the previous sections in order to characterize disordered textures. Many more measures and experiments are possible, but we have chosen a few to give the reader a flavor for the techniques used to analyze disordered textures.

5.4.1   IMPLEMENTATION RESULTS FOR COMPUTING THE ENTROPY AS A MEASURE FOR DISORDER

The above method was implemented for four co-occurrence matrices with $d = 1$ and θ corresponding to the horizontal, vertical, right-diagonal and left-diagonal directions. No distinction was made between left-to-right and right-to-left transitions. The measures for entropy derived from each co-ocurrence matrix were averaged to yield the final result. Results are displayed in table 5.1.

5.4.2   USING THE EXPECTED VALUES OF INTENSITY DIFFERENCES

The method described in section 5.3.1 was used to estimate the fractal dimensions of surfaces of differing roughness. Figure 5.2 shows the graph used

Results for computing the entropy derived from the cooccurrence matrix	
Image name	Entropy
birch	0.627279
d19	1.313577
fract	1.435938
d9	1.470659
d4	1.509898
d2	1.590515
d5	1.744302
d12	1.796189

TABLE 5.1. Computing the entropy of the texture

to perform linear regression to estimate H. Two representative images have been chosen to illustrate the nature of the plots. The correlation coefficient is a measure for how good the linear fit is to the data, and can be used to guage the applicability of the fractal model to the given image. A poor correlation coefficient will indicate that the fractal model is unsuitable for the given data.

Figure 5.3 shows several test images to which this method was applied. The results are summarized in table 5.2. The results have been arranged in increasing order of fractal dimension to facilitate comparison.

5.4.3 RETICULAR CELL COUNTING METHOD FOR THE FRACTAL DIMENSION

The method described in section 5.3.2 was used to estimate the fractal dimensions of surfaces of differing roughness. Figure 5.4 shows the graph used to perform linear regression to estimate D in equation 5.6. The two representative images selected in the earlier section have been chosen to illustrate the nature of the plots. Over the range of scales shown, the plot is linear, showing that the fractal model is appropriate to describe the surface.

Table 5.3 tabulates the results obtained for the rest of the images shown in figure 5.3.

As an illustration of the fractal dimension as a measure of surface roughness, we have applied the method to quantify surface roughness in integrated circuit images. Such measures are useful for quality control, as described in [104]. The results of applying the reticular cell counting method to some IC images are shown in figures 5.5(a) through (d). The boxes in the image show the regions selected for roughness measurement. The numbers

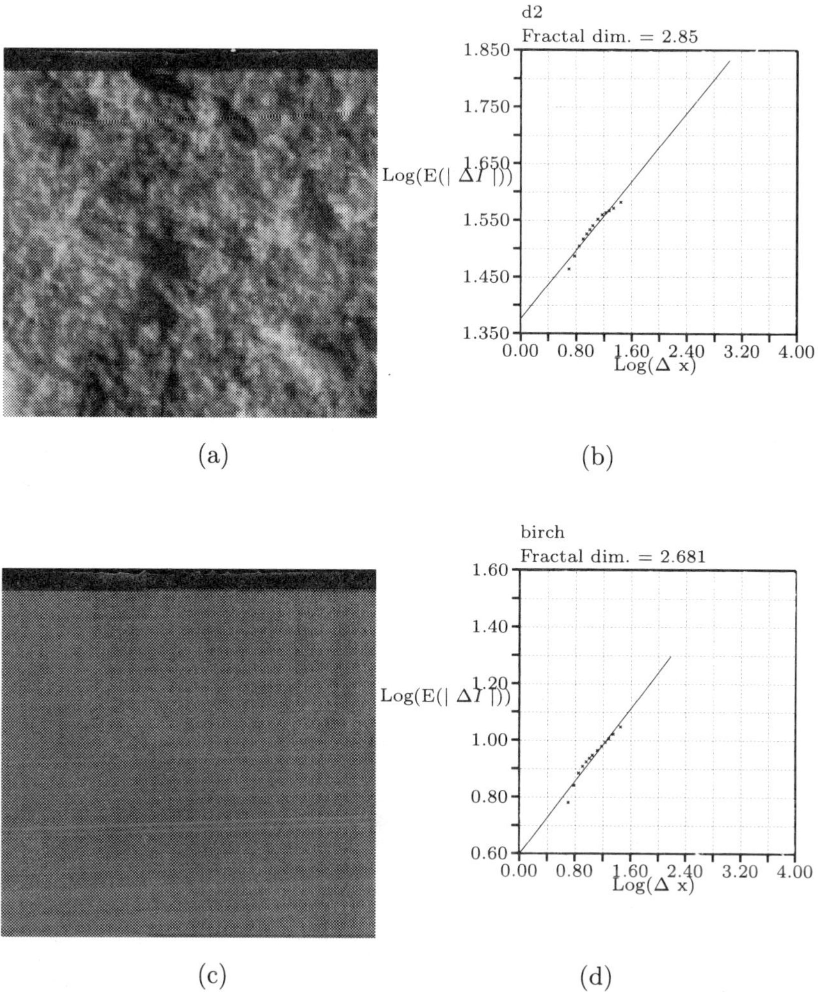

(a)   (b)

(c)   (d)

FIGURE 5.2. (a) The original image, consisting of fieldstone from [19] (b) The linear regression plot for estimating the fractal dimension, as derived from equation 5.5. The data points are indicated by crosses, and the solid line represents the least squares fit. The slope of the line is H, and is used to determine the fractal dimension according to equation 5.3 (c) The original image, consisting of a sample of birch wood, from [20] (d) The linear regression plot for estimating the fractal dimension, as derived from equation 5.5.

d4

d5

d9

d12

FIGURE 5.3. Images used to test the algorithms for fractal dimension estimation. d4, d5, d9, d12, were obtained from [19].

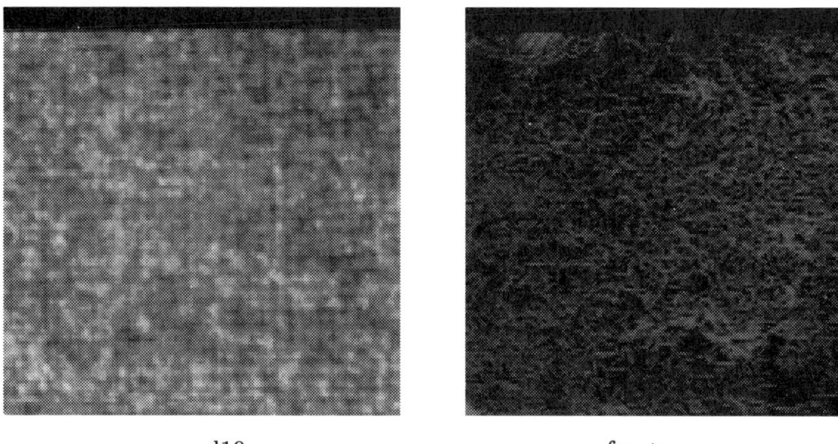

FIGURE 5.3. Images used to test the algorithms for fractal dimension estimation (contd.) d19 was obtained from [19]. 'fract' depicts a fractograph specimen from [137]. (Reproduced with permission from *Metals Handbook*, Vol 9, 8th Ed., *Fractography and Atlas of Fractographs*, American Society for Metals)

Results for computing the fractal dimension using the expected values of intensity differences		
Image name	Fractal dimension	Correlation coefficient
birch	2.6808	0.972137
d2	2.84964	0.971721
d12	2.87725	0.997946
d19	2.89358	0.968746
d4	2.90558	0.987498
d9	2.90983	0.992363
fract	2.92754	0.989570
d5	2.93932	0.957704

TABLE 5.2. Computing the fractal dimension by using the expected values of intensity differences

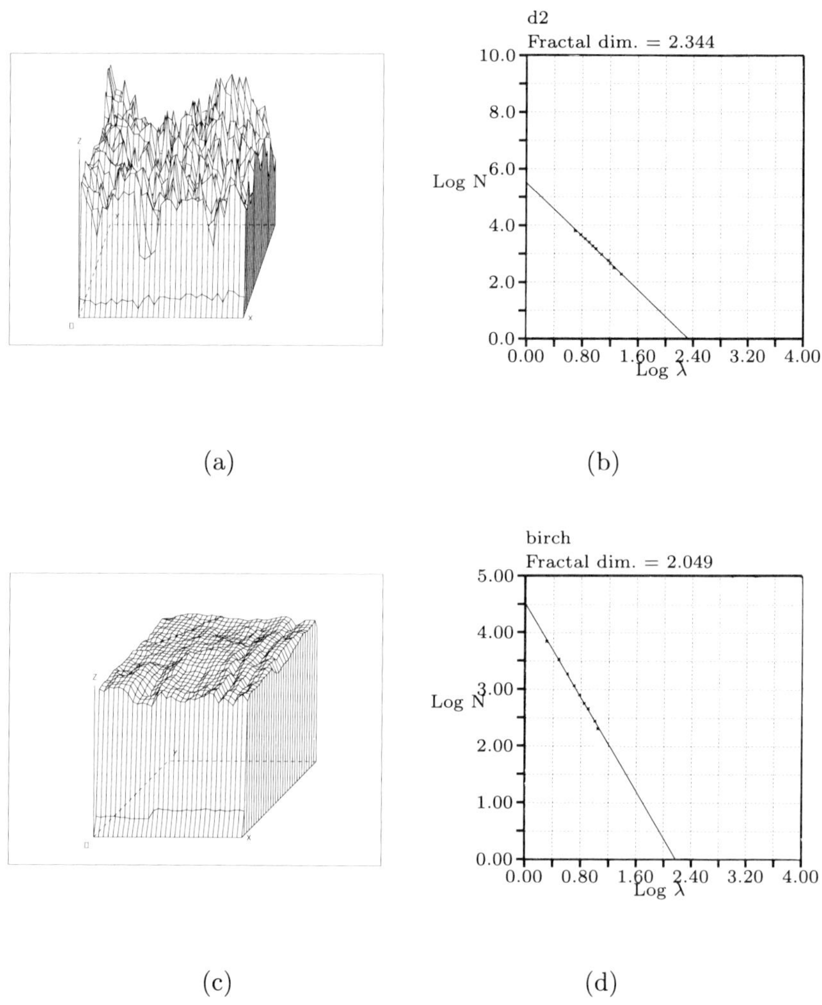

FIGURE 5.4. Computing the fractal dimension using the reticular cell counting approach. (a) Analysis of the original image, consisting of fieldstone from [19] This plot shows the 3D intensity surface. (b) The linear regression plot for estimating the fractal dimension, as derived from equation 5.6. The data points are indicated by crosses, and the solid line represents the least squares fit. The slope of the line is $-D$, where D is the fractal dimension. (c) Analysis of the original image, consisting of a sample of birch wood from [20] This plot shows the 3D intensity surface. (d) The linear regression plot for estimating the fractal dimension, as derived from equation 5.6.

Results for computing the fractal dimension using the reticular cell counting method		
Image name	Fractal dimension	Correlation coefficient
birch	2.049	0.9993
d19	2.32577	0.99921
d2	2.34406	0.99846
d9	2.39479	0.99920
d5	2.39554	0.99718
d12	2.42113	0.99762
d4	2.4229	0.99906
fract	2.487729	0.9995

TABLE 5.3. Computing the fractal dimension by the reticular cell counting method

within the boxes show the fractal dimension computed for that part of the image.

5.4.4 DISCUSSION OF RESULTS

1. As one can see from table 5.1, these results are quite different from those provided by using the fractal dimension as a measure of surface roughness, indicating that entropy does not correlate well with a human's assessment of surface roughness.

2. There are also many problems associated with the use of the cooccurrence matrices. Firstly, a cooccurrence matrix has to be computed for different values of d and θ, and then the appropriate texture measure has to be computed from each cooccurrence matrix.

   Secondly, let us look at the time complexity of the algorithm. Assume that we have G gray levels in the image, the image size is NxN, and the computation is carried out over M scales, and all discrete angles θ that are permissible be the choice of the scale d. For a given scale d, we have $4d$ possible values for θ. The time complexity of this algorithm is

$$\sum_{d=1}^{d=M} 4d(N^2 + G^2) = O(M^2 N^2 + M^2 G^2) \qquad (5.12)$$

   If we are to use different scales d, then this method becomes computationally more expensive than the fractal method.

3. There appears to be a disagreement between the results obtained using the two different techniques for fractal dimension measurement.

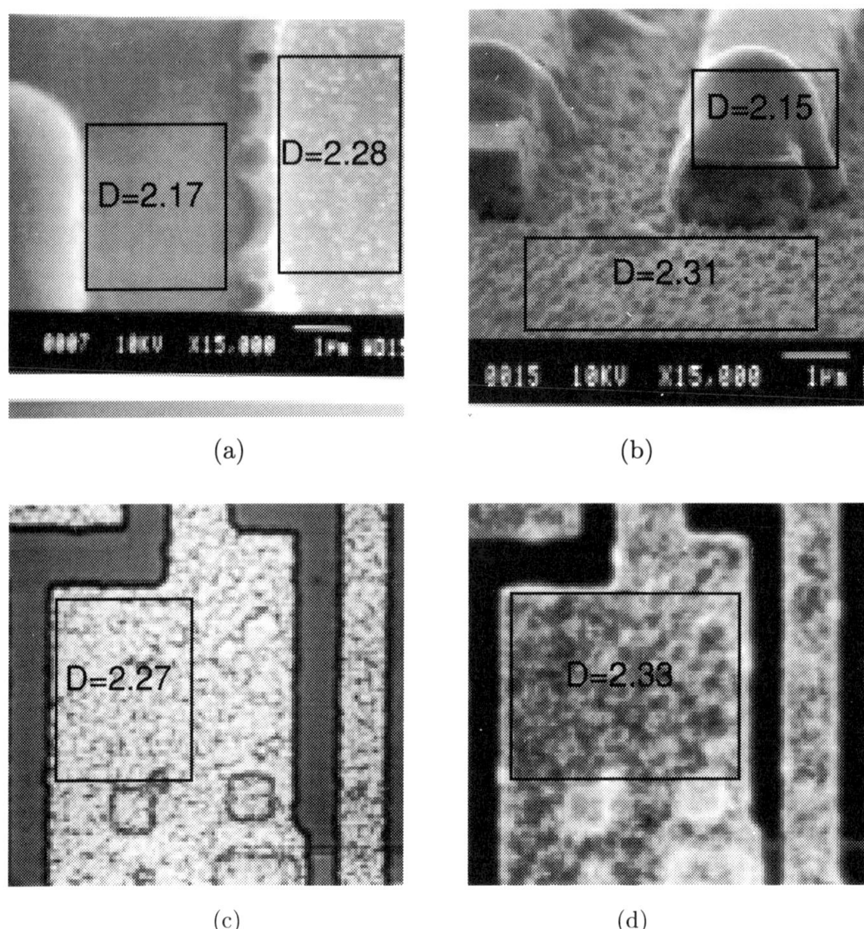

(a)   (b)

(c)   (d)

FIGURE 5.5. The results for surface roughness measurement using fractal dimension. (a): Shows a SEM image of a silicon wafer, consisting of an oxide surface (which should be optically smooth and flat), and aluminium (evaporated). The surface on the right is rougher than the surface on the left, hence the fractal dimension for the right surface is higher than that of the left surface. This result shows that the fractal dimension corresponds to the intuitive ordering of surface roughness. (b): Shows a SEM image of a milled GaAs surface and metal line. The surface consists of metal deposited via evaporation on a substrate after an RIE etch process. (c): Shows an image of a silicon wafer obtained through an optical microscope. This image consists of (evaporated) aluminium on the wafer. (d): Shows a dark field image of the same silicon wafer.

This is due to the fact that different models are being used to describe the fractal surfaces, as indicated by equations 5.2 and 5.6. However, this does not alter very much the rank ordering of the images according to fractal dimension – qualitative trends are maintained.

5.5   Conclusion

In this chapter we briefly examined a few techniques used to describe disordered textures. We feel that many of the methods present in the literature can be successfully used for the analysis of disordered textures.

We have demonstrated the usefulness of the fractal dimension as a surface roughness measure by analyzing images from Brodatz [19], as well as images from a real world application – that of semiconductor wafer inspection.

6

Compositional textures

6.1 Introduction

Color is characterized in terms of brightness, hue and saturation of the RGB components [128]. Due to this three dimensional nature of color, traditional color specification systems use a triple of numbers to describe color (see [48, 34] for a discussion of conversions between various systems). Unfortunately, such a parallel does not exist for texture – we do not know what the dimensions of texture are. A possible insight could be gained by analyzing the classes of texture presented in the earlier chapters and finding out ways in which these textures can be combined. The aim of this chapter is to define such compositional schemes.

By using these compositional schemes, we complete the taxonomy for texture that we seek. The last class of textures we consider in this book are those of *compositional textures*. Compositional textures are created from the classes of disordered, strongly ordered and weakly ordered textures analyzed earlier, according to compositional rules to be presented in this chapter.   The rules of composition allow the construction of arbitrary textures, and in conjunction with the models for different kinds of textures, form a representational scheme for texture.

Once we have such a representation for texture, it will suggest techniques to extract a symbolic structure from a given texture. Of course, in the transformation from the iconic to the symbolic, certain information is lost, and we would like to ensure that the descriptive power of the symbolic representation scheme is adequate. The question is: how can the symbolic representation scheme be evaluated?

One way of evaluating a representational scheme for texture is to test whether it can describe textures from some standard data set, or to test it on a sufficiently large sample of textures from the real world. The world of texture is vast, and even an album of textures such as Brodatz's [19] does not constitute a sufficient sample for the purpose of testing.

Nevertheless, our taxonomy for texture has been tried in a *gedanken* experiment fashion on several textures from Brodatz [19, 20], and Van Dyke [118], and appears to be capable of generating meaningful symbolic descriptions. In this chapter, we show how *in principle*, all the textures from Brodatz may be described using the taxonomy presented in this book. Though the complete implementation of the taxonomy falls outside the scope of the book, directions for such an effort are delineated in section 6.10.2.

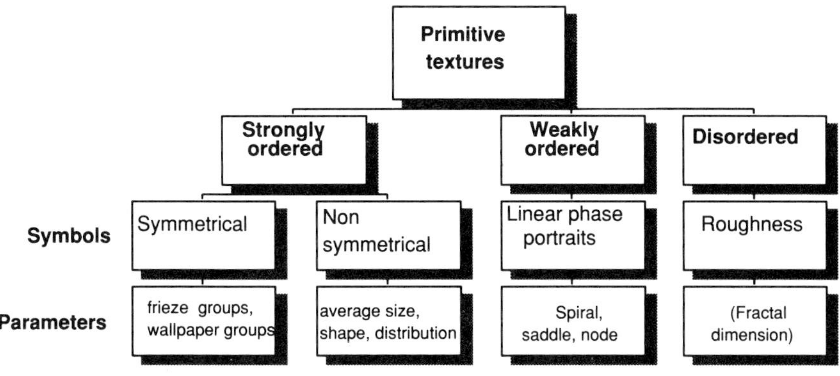

FIGURE 6.1. Illustrating a partial taxonomy for texture

6.2   Primitive textures

Thus far we have examined three different classes of texture in chapters 3, 4 and 5. These can be summarized in the form of a partial taxonomy as in figure 6.1 (the reason for calling it a partial taxonomy is that it can handle about 91% of the textures in Brodatz [19], as shown in table 6.1).

These three classes of texture, namely *strongly ordered, weakly ordered,* and *disordered* are collectively termed **primitive textures**. Examples of the different types of primitive textures are shown in figure 6.2. **Strongly ordered textures**, symbolized by **S**, are those which exhibit either a specific placement of some primitive elements (like a brick wall, figure 6.2a) or are comprised of a distribution of a class of elements (such as grains within a matrix, figure 6.2b). **Weakly ordered textures**, symbolized by **W**, are those that exhibit some degree of orientation specificity at each point of the texture (figure 6.2c). **Disordered textures**, symbolized by **W**, are those that show neither repetitiveness nor orientation and may be described on the basis of their roughness (figure 6.2d).

Table 6.1 shows a classification of about 91% of the textures from Brodatz's album *Texture* [19] according to the above partial taxonomy. In fact, another album by Brodatz, *Wood and wood grains* [20], contains several more examples of weakly ordered textures which have not been mentioned in the table. From this table it should be clear that it would be desirable to move towards the taxonomy described in figure 6.1. This would be capable of generating descriptions for a large number of textures, both natural and man-made.

An important question that must be addressed at this point is how can we account for the remaining 9% of the textures from Brodatz [19] that are

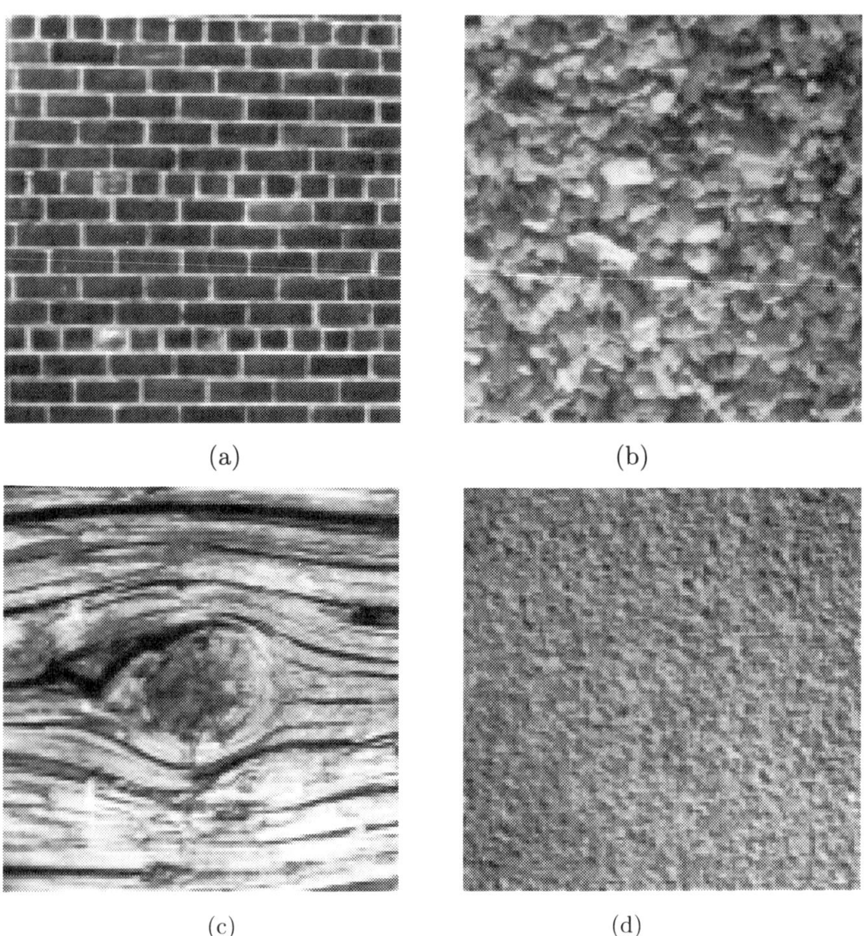

(a)   (b)

(c)   (d)

FIGURE 6.2. Illustrating the taxonomy for texture; **(a)** A brick wall, (d95 in Brodatz [19]). **(b)** Pebbles (d5) **(c)** Knot in a piece of wood (d72) **(d)** Surface of cork (d4)

not represented in table 6.1. Briefly, our approach is to create an additional class of textures, called *compositional textures*, which accounts for those textures that the partial taxonomy of figure 6.1 fails to classify. This is discussed in greater detail in section 6.6.

6.3   A Parametrized symbol set

So far we have presented a taxonomy for texture. A nice feature about this taxonomy is that it can be viewed as a parametrized symbol set Σ. Each member $\sigma_i \in \Sigma$ ($i \in [1..N]$), where N is the cardinality of Σ) has associated with it a set of parameters, p_i. These parameters depend on the particular symbol σ_i. Thus, we have $\sigma_i = \sigma_i(p_i) = \sigma_i(p_{i_1}, p_{i_2}, ...p_{i_{M_i}})$, where M_i is the number of parameters associated with the ith symbol.

Let $\mathbf{S}$ denote a strongly ordered texture, $\mathbf{W}$ a weakly ordered texture and $\mathbf{D}$ a disordered texture. We define a *primitive texture* to be one which can be entirely labeled as $\mathbf{S}$, $\mathbf{W}$, or $\mathbf{D}$. This implies that a primitive texture is homogeneous in the sense that it does not have textural discontinuities within it. Furthermore, there is only one *dominant* description for the given texture, which is $\mathbf{S}$, $\mathbf{W}$, or $\mathbf{D}$. The reason for introducing the dominance constraint is that a strongly ordered texture can also be considered to be weakly ordered in the following sense. Since the boundaries of the structural element define dominant local orientations, the strongly ordered texture containing these elements satisfies the definition of a weakly ordered texture given in section 6.2. Hence, it is necessary to select the most structurally complex description possible for a given texture.

Thus, complex textures such as the one to be discussed in figure 6.4 are not considered to be primitive textures. The reason for this will become clear in sections 6.6 and 6.7. This type of definition brings up the issue of how to decide whether an arbitrary texture is primitive – this will be discussed towards the end of the chapter.

Table 6.2 also summarizes the connection between a texture type, its symbolic representation and the underlying mathematical model. The mathematical model specifies what parameters are used to describe a given texture type. Thus, strongly ordered textures can be described by using the theory of symmetry, weakly ordered textures by using the geometric theory of differential equations, and disordered textures by fractal geometry.

As a concrete illustration of this parametrized symbol set, consider the symbols that describe oriented textures, as presented in chapter 3. In this chapter, we illustrated the effect of varying the parameters in the matrix $\mathbf{A}$ on the visual appearance of the flow. Figure 3.35(a) showed a spiral with divergence dominating. The matrix $\mathbf{A}$ in this case was $a_{11} = 1$, $a_{12} = -0.5$, $a_{21} = 0.5$, $a_{22} = 1$. Figure 3.35(b) showed a spiral with rotation dominating. The matrix $\mathbf{A}$ in this case was $a_{11} = 1$, $a_{12} = -2$, $a_{21} = 2$, $a_{22} = 1$. Thus,

Texture Type	Textures from Brodatz [19]
Disordered	d2, d7 (fieldstone), d4 (pressed cork), d9 (grass lawn), d57 (handmade paper), d86 (ceiling tile), d90, d91 (clouds), d92 (pigskin), d100 (ice crystals on an automobile)
Strongly ordered	d1, d6, d8, d14 (woven aluminium wire), d3, d10, d22, d35, d36(reptile skin), d5 (mica), d18 (raffia weave), d20, d21 (french canvas), d23, d27, d28, d29, d30, d31, d54, (beach pebbles), d25, d26, d94, d95, d96 (brick wall), d32, d33 (pressed cork), d34 (netting), d46, d47 (woven brass mesh), d48 (masonite panel), d49 (straw screening), d52, d53 (oriental straw cloth), d55, d56 (straw matting), d59, d62 (european marble), d64, d65 (handwoven oriental rattan), d66, d67 (plastic pellets), d73 (soap bubbles), d74, d75 (coffee beans), d77 (cotton canvas), d78, d80, d81, d82(oriental straw cloth), d79 (oriental grass fiber cloth), d83, d85 (woven matting), d84 (raffia), d88, d89(dried hop flowers), d98, d99(crushed rose quartz), d101, d102(cane), d103, d104(loose burlap), d105, d106(cheesecloth), d111, d112(plastic bubbles)
Weakly ordered	d12, d13 (bark of tree) d15 (straw), d16, d17 (herringbone weave), d24 (pressed calf leather), d37, d38 (water), d43, d44, d45 (swinging lights in a darkened room, d50, d51 (raffia), d68, d69, d70, d71 (wood grain), d72 (tree stump), d87 (sea fan), d93 (fur), d97 (tree stump), d107, d108, d109, d110 (paper)

TABLE 6.1. Classification of textures from Brodatz [19] into the classes of strongly ordered, weakly ordered and disordered. The numbers of the textures are indicated and their description is given in parantheses.

Texture Type	Symbolic description	Underlying mathematical model
Disordered	rough, smooth etc.	Fractal geometry
Strongly ordered	motif and placement rules	Theory of group symmetry
Weakly ordered	nodes, spirals, saddles	Geometric theory of differential equations

TABLE 6.2. Correspondence between texture types and mathematical models

though the symbolic description of the texture remains the same, the exact visual representation could vary.

The advantage of using a parametrized symbol set is that it allows both a qualitative and quantitative scheme for describing textures. If one omits a description of the parameters, ie specifies only the terms σ_i without mentioning p_i, then one has a purely qualitative description of the given texture. Thus one can talk about a texture being rough or grainy, without specifying the exact measure for the roughness. In several applications, a purely qualitative description like this is sufficient. For instance, in semiconductor wafer inspection, if one observes a pattern called star-burst, it is enough to suspect that some error occurred during the spinning process [35]. The exact nature of the star-burst pattern (ie its quantitative features) is irrelevant to the diagnostic process.

On the other hand, if one specifies the values of the parameters p_i, then it allows us to describe the texture quantitatively. For instance, if one specified the matrix $\mathbf{A}$ that corresponds to a linear flow that causes a spiral texture, one can reconstruct the original texture based on its parametrized representation. Other kinds of decisions can also be made: for instance in quality control, one can reject a certain part if the roughness parameter is beyond a certain limit.

6.4   Three types of composition

As mentioned in the earlier section, we select primitive textures to define compositional schemes. Since the texture has to be defined over some region $\mathbf{R}$ in space, assume the region $\mathbf{R}$ to be the unit square, without loss of generality. The first two compositional schemes deal with the unit square region. Arbitrarily shaped regions are handled by the third compositional scheme in section 6.7.

With these assumptions, we can now create more complex textures by using the techniques outlined in the following sections. Basically there are three types of composition, entitled linear combination, functional composition and opaque overlap. These three compositional techniques are capable of generating a wide variety of textured image.

6.5   Linear combination (transparent overlap)

This is the simplest combination scheme for textures. Let t_1 and t_2 be any two textured images. Then, the linear combination of these two textures results in a third texture t_3, as given by the following equation

$$t_3 = c_1 t_1 + c_2 t_2 \tag{6.1}$$

where c_1 and c_2 are real numbers.

The reason this is termed a transparent overlap is that one could consider the two textures to be drawn on transparent sheets, and we look the texture resulting when the two sheets are overlayed one on top of the other.

In practice, since image intensity values have to be positive, and are customarily quantized into 256 gray levels, one could take the resulting image in equation 6.1 and map it into the [0,255] range.

This method is illustrated by the examples in figure 6.3. Thus, one could have a realistic looking brick wall generated as the linear combination of a disordered texture (e.g. sand) and an ideal structural texture (e.g. a rectangular motif replicated according to the sides of a parallelogram).

In general, it is difficult to predict the nature of the resulting texture when t_1 and t_2 are drawn from the sets $\mathbf{S}$, $\mathbf{W}$ and $\mathbf{D}$. As shown in the case of figure 6.3, the linear combination of a strongly ordered texture with a disordered texture could give either another strongly ordered texture or a disordered texture. Hence $\mathbf{S} + \mathbf{D} \longrightarrow \mathbf{S}$ or $\mathbf{S} + \mathbf{D} \longrightarrow \mathbf{D}$. Thus, this does not allow one to create a simple composition table for the operator '+'.

Another interesting issue is that a given texture can be created in a variety of different ways, and it may not be possible to single out a unique representation. For instance, a brick wall pattern can be created by replicating a rectangle in a regular fashion. Alternately, it can be created by linearly combining two oriented textures, where the first texture is composed of vertical lines and the second is composed of horizontal lines.

6.6   Functional composition

This type of composition is more complex than the linear combination presented in the previous section. The essential idea is to embody the features of one class of textures within the framework of another class of textures. For instance, one could embody a structural texture such as a brick wall in the framework of a flow-like texture such as vortex flow – the resulting texture would appear as shown in figure 6.4. This kind of composition will be denoted by the operator '$\odot$'. Similar kinds of embodiment can be performed with the other classes of texture, resulting in the compositional definitions given below.

Note that different interpretations could be attached to the operator '$\odot$'. In this book, I have suggested one set of possible definitions for the operator '$\odot$'. Since we are considering three classes of texture, $\mathbf{S}$, $\mathbf{W}$ and $\mathbf{D}$, we have to give an interpretation for '$\odot$' in nine possible cases.

6.6.1 $\mathbf{W} \odot \mathbf{S}$

The meaning of the operator '$\odot$' in this case is as follows: take the strongly ordered texture and submit it to a coordinate transformation indicated by the weakly ordered texture. Such a situation is illustrated in figure 6.4,

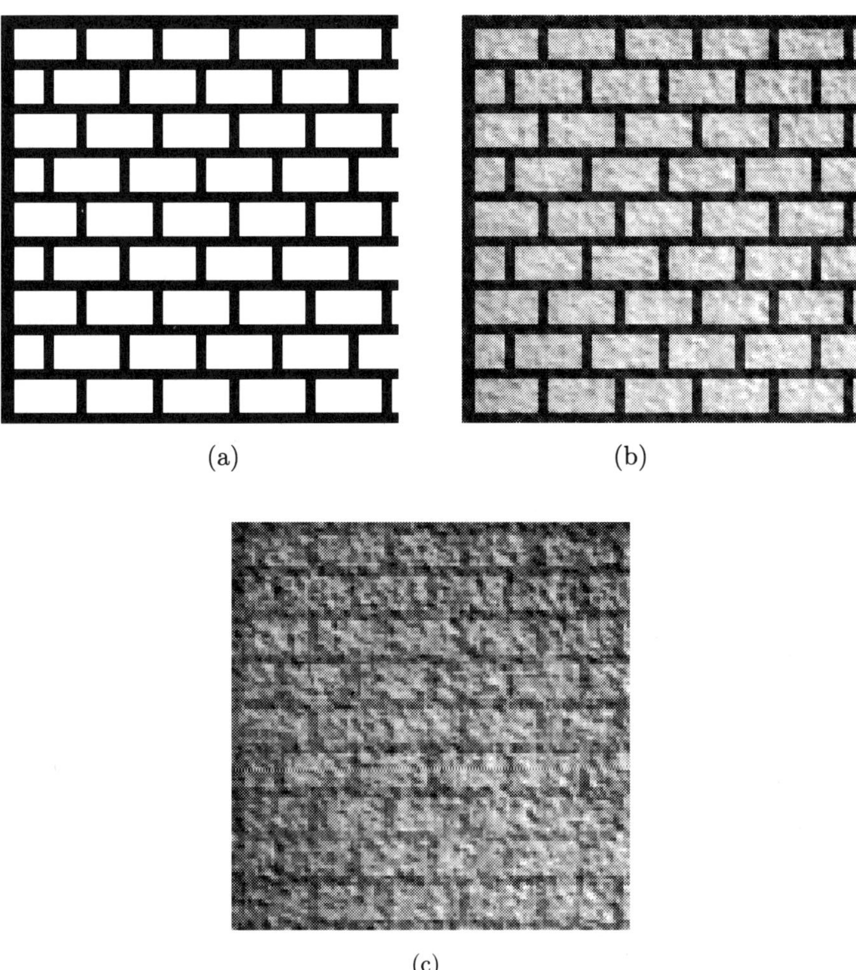

(a)

(b)

(c)

FIGURE 6.3. Illustrating linear combinations of textures. (a) An ideal repetitive pattern, eg. a brick wall. (b) The combination of the ideal repetitive pattern with a disordered texture, specifically the texture in figure 6.2(d). The weights for the two textures are 1 for the repetitive pattern and 1.5 for the disordered texture. (c) The linear combination when the weights are 1 for the repetitive pattern and 10 for the disordered texture. Note that the resulting texture does not retain any structural information.

where the strongly ordered texture is distorted according to the weakly ordered texture.

This kind of operation has a flavor similar to the work done by Kass and Witkin [65]. In this paper, the authors first describe a method to extract dominant local orientation, resulting in a flow field. The flow field is then used to extract a coordinate system in which to view the pattern. The coordinate system consists of integral curves along the direction of flow, and an orthogonal family across the direction of flow. Viewing the image in this coordinate system reveals the underlying pattern independent of orientation distortions. Thus, they are able to take a wood knot, or a piece of burlap and unwarp the distortion that caused the oriented texture to occur.

A similar type of analysis could in principle be applied to the figures in 6.4, to result in a structural description. Thus, one would first unwarp the distortion due to spiral flow, resulting in a residual texture corresponding to a brick wall. The residual texture can then be described by a structural analysis technique.

A special case of a flow-like distortion is perspective distortion. This case has been considered by Hamey [43] in his thesis, where he extracts locally repetitive structure in an image of a skyscraper seen from the top. This image is the result of a perspective distortion. Hence, it seems that a direct structural approach such as Hamey's can handle simple kinds of distortion. However, for complex cases such as the one shown in figure 6.4, a direct structural approach will fail.

A situation similar to perspective distortion is described by Lockwood and Macmillan [75], where the term dilation symmetry is used for this case. Also, patterns exhibiting spiral symmetry are discussed in this book.

6.6.2   D $\odot$ S

In order to give such an operation meaning, one would model the disordered texture as a set of randomly placed points on the plane. The structuring element of the strongly ordered texture could then follow the placements indicated by the points of the above disordered texture.

Such an approach has been considered by Ahuja and Rosenfeld [1]. They propose a class of image models, called mosaic models, which are based on random geometric processes. A *coverage mosaic* is obtained by a random arrangement of a set of geometric figures in the plane. The random arrangement is modeled by a Poisson point process which drops the centers of the figures onto the plane, and the figures considered are line segments, ellipses, circles, rectangles and squares.

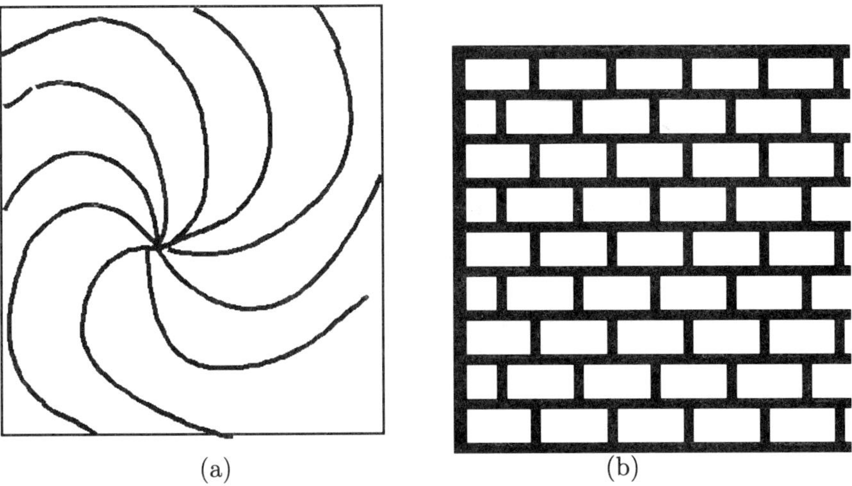

(a)

(b)

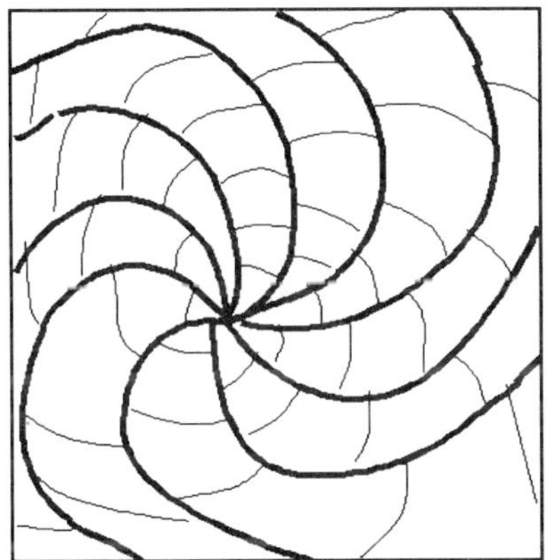

(c)

FIGURE 6.4. The result of composition of an oriented texture with a strongly ordered texture. (a) Shows a spiral flow-like texture (b) Shows a structural texture with rectangular elements. (c) Shows the result of subjecting the structural texture through the distortion indicated by the spiral texture.

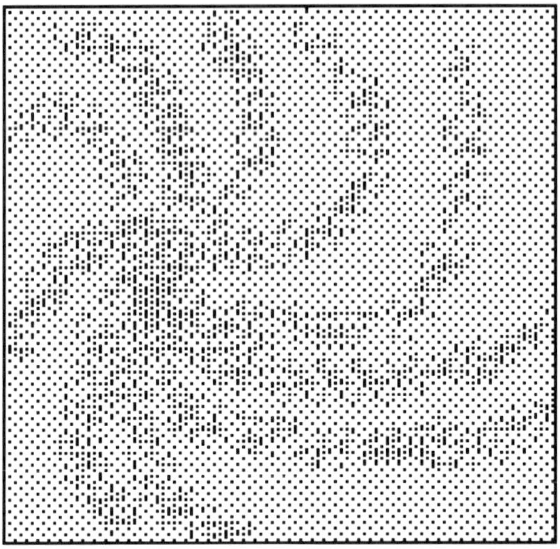

FIGURE 6.5. Illustrating the distortion of a disordered texture according to a weakly ordered texture

6.6.3 $\mathbf{D} \odot \mathbf{W}$

This case is more difficult to give an interpretation to. If the earlier bombing model of Ahuja and Rosenfeld is extended, then this composition could be taken to mean a random distribution of point singularities of the orientation field of W, the weakly ordered texture. Thus, one could consider the orientation field around a local neighborhood of a singular point, and distribute such singular points randomly in the plane.

6.6.4 $\mathbf{W} \odot \mathbf{D}$

In this case, we take a disordered texture D, and subject it to a warping indicated by W. Figure 6.5 shows a spiral organization of clusters of dots. This is an example of distorting a disordered texture according to another flow-like texture.

Thus, in order to describe such a texture, one would have to unwarp it, and describe the statistics of the residual texture using a few parameters. This sort of analysis has been proposed by Kass and Witkin [65]. In this paper, the authors mention that the results of this technique are forthcoming; however, this work does not appear to be published.

6.6.5 $\mathbf{S} \odot \mathbf{D}$

In this case, we can consider the structuring element or motif of S to be made up of a texture from D. Thus, one could for instance have a rectangular element containing a rough texture of sand. If this rectangular element is replicated according to any of the wallpaper patterns in section 4.6, then we have an image of a realistic looking brick wall.

Note that this composition is subtly different from a simple linear combination of the two textures. In the linear combination case, the description of the disordered texture can vary across the image, but in the case of the operator '$\odot$', the disordered texture within the motif remains constant throughout the image.

6.6.6 $\mathbf{S} \odot \mathbf{W}$

This case is similar to the previous one, where the motif contains some arbitrary flow-like texture. As an example, wooden floorings have such a description.

6.6.7 $\mathbf{S} \odot \mathbf{S}, \mathbf{D} \odot \mathbf{D}$ and $\mathbf{W} \odot \mathbf{W}$

Finally we are left with the functional composition of two textures belonging to the same class. The interpretations given to the operator '$\odot$' in the earlier sections can be used in these cases.

Thus, $\mathbf{S} \odot \mathbf{S}$ means the following: take a structural texture as a motif, and then replicate this motif according to given placement rules.

$\mathbf{W} \odot \mathbf{W}$ is interpreted as starting with a given oriented texture and distorting it according to a second oriented texture.

Finally, $\mathbf{D} \odot \mathbf{D}$ is interpreted as starting with a random group of points and then replicating this group in a random fashion in the plane.

The analysis of these cases completes the interpretation of the operator '$\odot$' for the three classes of texture $\mathbf{S}, \mathbf{D}$ and $\mathbf{W}$.

6.7   Opaque overlap

A texture mosaic is defined to be an image which has at least two different types of texture present within it – ie there is a texture boundary within the image. Let t be any texture created by the application of the above rules composed an arbitrary number of times with each other. In order to create an arbitrary texture mosaic, we first need to define the restriction of the texture t to a region $\mathbf{R}$ on the $x - y$ plane. A region $\mathbf{R}$ is taken to be the set of points in the plane which are enclosed by a simple closed curve. If the texture t is assumed to extend infinitely along the plane, then its restriction to $\mathbf{R}$ is simply the texture that falls within $\mathbf{R}$ (such a definition is similar to the clipping window in computer graphics). This will be denoted by $t/\mathbf{R}$.

Now consider an ordered list of such restrictions, $(t_1/\mathbf{R_1}, t_2/\mathbf{R_2}, t_3/\mathbf{R_3},$ ... $t_n/\mathbf{R_n})$ Region $\mathbf{R_i}$ is said to obscure region $\mathbf{R_j}$ if $i > j$ in the above

FIGURE 6.6. Illustrating the formation of a texture mosaic from constituent textures generated by the compositional rules in sections 6.5 and 6.6

ordered list. A texture mosaic can then be composed from the above ordered list, if one follows the constraint of obscuring. In other words, one overlays the regions $\mathbf{R_1}$ through $\mathbf{R_n}$ such that the most recent region covers all other intersecting regions in an opaque fashion. Hence this compositional scheme is called an *opaque overlap*. Figure 6.6 illustrates such a mosaic, and will be discussed in more detail in section 6.8.

An implementation of this type of opaque overlap is available in packages such as *POSTSCRIPT*, [124] in terms of the `clippath` operator, and the way the pen is handled.

6.8 Definition of texture

We are now in a position to give a constructive definition for a textured image. A textured image is any image that can be created by either a linear combination or functional composition or both, followed by an opaque overlap, of primitive textures. A primitive texture is either **S**, **W**, or **D**.

Figure 6.6 shows the creation of a textured image using primitive textures and the composition rules. This texture mosaic was created by scanning texture images from Brodatz [19], subjecting individual images to perspective distortion, and then overlaying the distorted textures. The following textures from the album were used: *brick wall* (d95) for the two walls, *grass*

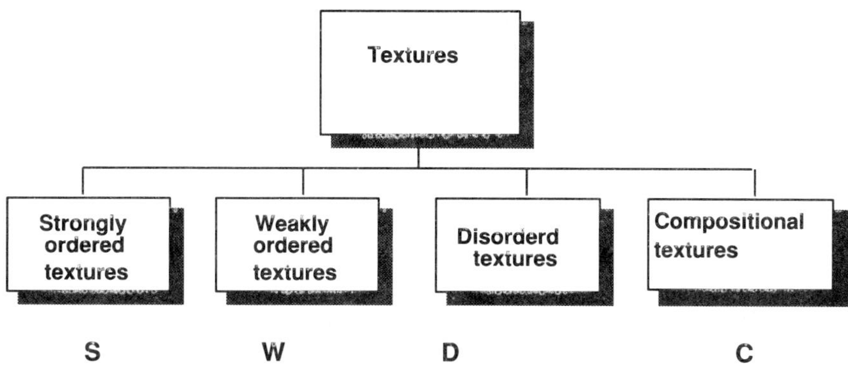

FIGURE 6.7. Members of the set of all textures

lawn (d9) for the floor, *wood grain* (d69) for the table, and genuine *reptile skin* (d22) for the chair. Note that the brick wall and reptile skin constitute strongly ordered textures, the wood grain on the table is a weakly ordered texture and the grass lawn is a disordered texture.

The distortion and overlaying was done using the application software called *Photoshop* on a Macintosh-IIX. This example shows that realistic scenes may be described in terms of the texture definition given above.

6.9   A complete taxonomy for texture

Having presented techniques for combining primitive textures, we are now in a position to create a complete taxonomy for texture. This is done by appending the class of compositional textures, C, to the partial taxonomy of figure 6.1. Figure 6.7 shows the final taxonomy, and identifies the symbols we will be using to describe the set T of textures. An element of T belongs to one of the following sets: S, W, D, C. where C denotes the set of compositional textures. Membership is defined in the following manner. $t \in T$ is a member of S if t is a strongly ordered texture, and so on.

The effect of introducing the class of compositional textures is to account for those textures that could not be handled by the partial taxonomy. Figure 6.8 illustrates some of the textures from Brodatz that fall into the category of compositional textures.

Figure 6.8(a) shows an image of a lace (texture d41) that is really a compositional texture. This can be created by filling the outline of the flowers with one kind of strongly ordered textures, and using this to opaque overlap a background of another kind of strongly ordered texture. Thus, we are using the compositional scheme of opaque overlap here to create the composite lace texture. Textures d39, d40 and d42 can be created in a similar fashion.

Texture Type	Relevant operator
Disordered	Statistical ($\mathcal{D}$)
Strongly ordered	Structural ($\mathcal{S}$)
Weakly ordered	Orientation field analysis ($\mathcal{W}$)

TABLE 6.3. Correspondence between texture types and operators

Figure 6.8(b) illustrates a texture of European marble (d61), which can be considered to be an overlap of a strongly ordered texture over a background of a disordered texture. Figure 6.8(c) shows another texture of marble (d63), which can be created by the composition of a weakly ordered texture and a disordered texture. The reason for introducing a weakly ordered texture into the description is that the original texture exhibits streaks in different directions. Finally, figure 6.8(d) shows a texture of woolen cloth (d19), which can be considered to be a linear combination of a disordered texture (representing the soft tufts) and a strongly ordered texture (representing the weave of the cloth). The area of petrology [78] abounds in many complex compositional textures, of which Brodatz has illustrated the instances of European marble.

Hence, if figure 6.7 is used to classify the textures in Brodatz, then this would result in the classification shown in figure 6.9.

6.10   Implementing the taxonomy

Though the implementation of an algorithm that would automatically perform a classification such as in figure 6.9 is very desirable, it is beyond the scope of this book. At this point, we can only outline a possible direction that seems plausible. Such a direction is briefly sketched below.

6.10.1   OPERATORS FOR TEXTURE

Let us consider the primitive textures presented earlier. Another way of looking at the work presented in chapters 3, 4 and 5, is to consider it as defining a set of operators. An operator acts on a given texture and produces both a qualitative and quantitative description (symbols and parameters). Thus, each type of texture is associated with an operator that is *most relevant* to it. By 'relevance' we mean that one and only one description **(D or W or S)** is best suited for a given uniform texture. The relevant operator is then associated with the corresponding description. Table 6.3 gives the correspondence between a texture type and its relevant operator.

Thus, disordered textures are best described by statistical techniques and

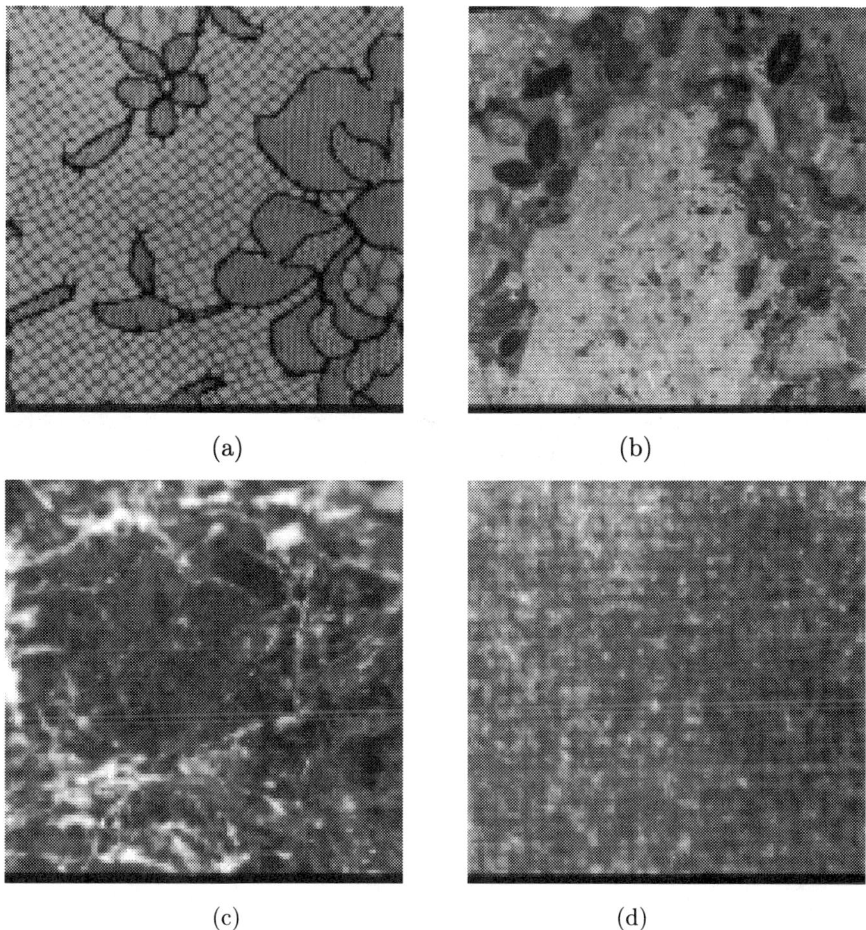

(a)

(b)

(c)

(d)

FIGURE 6.8. Illustrating some compositional textures from Brodatz. (a) An image of lace, showing fine stitch detail, (d41 in Brodatz [19]). (b) European marble (d61) (c) European marble (d63) (d) Woolen cloth, loosely woven with soft tufts (d19).

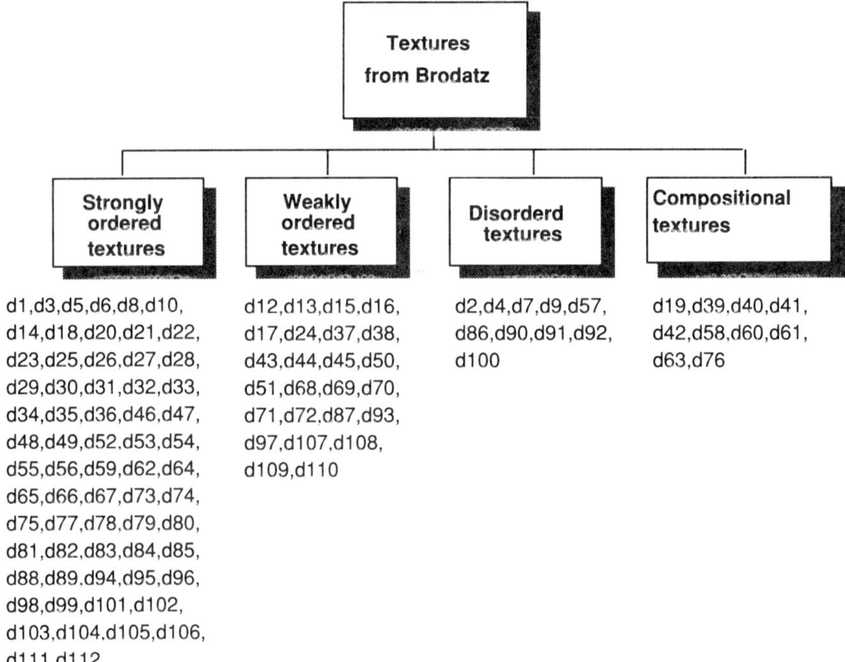

d1,d3,d5,d6,d8,d10,
d14,d18,d20,d21,d22,
d23,d25,d26,d27,d28,
d29,d30,d31,d32,d33,
d34,d35,d36,d46,d47,
d48,d49,d52,d53,d54,
d55,d56,d59,d62,d64,
d65,d66,d67,d73,d74,
d75,d77,d78,d79,d80,
d81,d82,d83,d84,d85,
d88,d89,d94,d95,d96,
d98,d99,d101,d102,
d103,d104,d105,d106,
d111,d112

d12,d13,d15,d16,
d17,d24,d37,d38,
d43,d44,d45,d50,
d51,d68,d69,d70,
d71,d72,d87,d93,
d97,d107,d108,
d109,d110

d2,d4,d7,d9,d57,
d86,d90,d91,d92,
d100

d19,d39,d40,d41,
d42,d58,d60,d61,
d63,d76

FIGURE 6.9. A classification of all the textures from Brodatz [19], using the taxonomy presented in figure 6.7

not by structural or flow techniques. (Note: at this stage we consider only uniformly textured regions. This is a reasonable assumption as there are several algorithms available to segment a texture mosaic into its uniformly textured components [79, 122].)

6.10.2 SEQUENCING OF THE OPERATORS

Given the availability of the above operators for texture, the description of an arbitrary texture could be obtained in principle by some sequencing of these operators. Now the issue is – what sequence of operators must we apply to a given texture. This kind of viewpoint leads to the interpretation of texture as consisting of a sequence of operator applications, where the operators are defined in table 6.3.

The goal of such an analysis should be to ultimately derive the *most ordered* description for the texture possible. Thus, a weakly ordered texture has orientation specificity at every point but must not exhibit structural groupings of elements. The reason for doing this is that a strongly ordered texture (such as a brick wall) is trivially weakly ordered. Similarly, one could have a brick wall, where the bricks are themselves disordered textures (i.e. a rough bricks). However, we are concerned with the highest level of description possible, i.e. the repetitive structure of the bricks.

Hence it would be tempting to impose the following hierarchy on the operators $\mathcal{D}, \mathcal{W}$, and $\mathcal{S}$: if $\mathcal{S}$ fails, use $\mathcal{W}$; if $\mathcal{W}$ fails, use $\mathcal{D}$. But the interesting question is – is the ordering of these operators fixed, or does it depend on the texture being analyzed? In most cases the sequencing $\mathcal{S} \longrightarrow \mathcal{W} \longrightarrow \mathcal{D}$ (apply $\mathcal{S}$ – if it fails, try $\mathcal{W}$ then $\mathcal{D}$) seems to work. For instance, many of the textures in Brodatz fall into this paradigm. However, for the case in 6.6.1, the operators must be sequenced as $\mathcal{W} \longrightarrow \mathcal{S}$.

The reason for this problem is that the primitive textures could be intermeshed in a very complex manner according to the compositional rules present in the earlier sections. The issue now is to devise a scheme that unravels the compositional structure of the texture as specified earlier. Unfortunately, we cannot answer this question at this point. A direction for such work is specified as follows. First segment the texture into homogeneous regions using an algorithm such as the one developed by Voorhees and Poggio [122] or Malik and Perona [79]. Then to each homogeneous region, do the following. Apply the oriented texture algorithm of Rao and Schunck [106]. Extract the local coordinate system from this orientation field as described by Kass and Witkin [65]. Then apply Hamey's algorithm [43] to detect repetition in the residual texture described in the local coordinate system.

Thus the issue of interpretation of an arbitrarily complex texture still remains an open issue. However, the direction specified in this book is one possible approach that could be tried.

6.11 Conclusion

In this chapter we presented a taxonomy for texture consisting of symbolic descriptions derived from germane mathematical models. These models are capable of describing disordered, strongly ordered and weakly ordered textures, which we term *primitive textures*.

We then presented a set of compositional rules that would allow the formation of more complex textures from these primitives. These compositional rules can be classified into the categories of functional composition, linear combination and opaque overlaps. Furthermore, the compositional rules define a new class of textures called compositional textures.

This approach defines a representation for an arbitrary texture in terms of primitive textures and composition rules. We showed in this chapter that in principle, all the textures in Brodatz's album [19] can be classified according to the taxonomy. What remains to be done is to automatically generate the structure of complex textures in terms of constituent primitive textures and compositions ie, what composition rules caused the texture to be generated.

Based on the ability of the taxonomy to classify the Brodatz textures [19], it appears that the methods described in this book are capable of capturing the essence of texture. This could be potentially useful in identifying the elements of texture from a computer graphics point of view, as these elements could be used to create realistic user-controlled textures.

The development of such a representation for texture forms an important step towards the creation of a visual language [104]. A visual language will be necessary to create and understand messages in the visual idiom. The significance of designing a visual language is that it will facilitate and standardize interactions and act as a vehicle for communication between the vision system and its external environment (users and machines). Naturally, the creation of a visual language requires a detailed analysis of the individual components of the visual process. Through the analysis of texture in this book, we have taken a step towards achieving this goal.

7

Conclusion

7.1 Summary of results

Texture is an important element of vision, and occurs in a variety of natural and man-made objects. The characterization of texture is becoming increasingly important as a diagnostic tool in advanced manufacturing applications.

In this book, we approached the problem of analyzing texture from the viewpoint of signal-to-symbol transformation. Though signal-to-symbol transformation is one of the central paradigms of computer vision, it has not been successfully applied to the visual cue of texture thus far. The reason for this is partly because there is no obvious choice of a symbol set that can describe texture. Thus, the unifying theme of this book has been to focus on a symbolic description scheme for texture.

The symbolic description scheme consists of a novel taxonomy for textures, and is based on appropriate mathematical models for different kinds of textures. The taxonomy classifies textures into the broad classes of disordered, strongly ordered, weakly ordered and compositional. Disordered textures are described by statistical measures, strongly ordered textures by the placement of primitives, and weakly ordered textures by an orientation field. Compositional textures are created from these three classes of textures by using certain rules of composition. The book strives to provide standardized symbolic descriptions that serve as a descriptive vocabulary for textures.

We briefly summarize the results obtained in this book.

7.1.1 Analysis of weakly ordered textures

Most of the experimental work in this book has focussed on weakly ordered, or flow-like textures. We presented a new algorithm for computing the orientation field for weakly ordered textures. The basic idea behind the algorithm is to use an oriented filter, namely the gradient of Gaussian, and perform manipulations on the resulting gradient vector field. We have proved the optimality of this algorithm in estimating the orientation direction. An added strength of the algorithm is that it is simpler and has a better signal to noise ratio than previous approaches, because it employs fewer derivative operations. We have also proposed a new measure of coherence, which works better than previous measures.

The angle and coherence intrinsic images represent a necessary initial stage in the processing of oriented textures. We justified the role that the texture orientation intrinsic images play by providing results from a number of experiments. These results indicate that operations on the angle and coherence images yield useful results in many domains. Given this important role that the texture orientation intrinsic images play, our results suggest the best way to compute these images.

The symbolic description of orientation fields is accomplished by using the geometric theory of differential equations and is based on the visual appearance of phase portraits. We employ a non-linear least squares technique to determine the phase portraits that best describe local regions of the orientation field. We have presented results of applying this technique to several real texture images. Specifically, we have segmented the given texture, derive its symbolic representation, and performed a quantitative reconstruction of the salient features of the original texture based on the symbolic descriptors. These results show that it is possible for a machine to automatically synthesize descriptions of complex oriented patterns. We have demonstrated the usefulness of this scheme by applying it to complex flow visualization pictures and defects in lumber processing.

7.1.2 STRONGLY ORDERED TEXTURES

Though repetitive textures have been analyzed by researchers in computer vision for a decade, few attempt to provide precise symbolic descriptions. In keeping with the spirit of this book, we suggest the use of well known techniques in the area of group theory and symmetry to aid the description of strongly ordered textures. The symbolic description schemes that we have in mind are those of the frieze patterns and wallpaper groups, which constitute a set of equivalence classes for strongly ordered textures.

The implementation of this idea is beyond the scope of this book, and is suggested as a future project.

7.1.3 DISORDERED TEXTURES

Since most of the effort in computer vision related to texture has focussed on the statistical approach, this topic is not dealt with extensively in the book.

We believe that fractal based techniques are the best way to describe disordered textures. The reason for this is the strong correlation between fractal measures for surface roughness and the subjective estimates provided by humans, as reported by Pentland [97].

7.1.4 COMPOSITIONAL TEXTURES

We have presented a set of compositional rules, consisting of linear combination, functional composition and opaque overlap, that allow the creation of more complex textures from the primitive textures belonging to the classes of strongly ordered, weakly ordered and disordered.

This approach defines a representation for an arbitrary texture in terms of primitive textures and composition rules. The development of such a representation for texture forms an important step towards the creation of a visual language [104]. Through the analysis of texture in this book, we have taken a step towards achieving this goal.

7.2   Contributions

The principal contributions of this book are the following:

- A taxonomy for the description of texture. This taxonomy provides standardized symbolic descriptors that can be used for qualitative descriptions of a wide variety of textures. The taxonomy is based on appropriate mathematical models for different kinds of textures, and can classify, in principle, all the textures from Brodatz [19]

- An important class of textures which has not received adequate attention in computer vision is that of flow-like or oriented textures. In this book, a complete signal-to-symbol transformation scheme for oriented textures has been designed and tested. This work establishes the importance of analyzing oriented textures, as they occur in a variety of applications. The algorithm for computing symbolic descriptions of oriented textures proceeds in two stages. The first stage consists of the computation of an orientation field, which abstracts dominant local orientation in the texture. The second stage uses linear phase portrait analysis to segment the orientation field, and derive symbolic descriptors.

- The strength of this book lies in applying these algorithms to several practical applications. Thus, examples are provided from the areas of semiconductor wafer inspection, fluid flow visualization and lumber Several other potential applications for these algorithms are also mentioned.

The additional contributions of this thesis are the following:

- This book tries to unify work done in several areas of texture analysis. This is done by proposing a symbolic description scheme for texture that employs some techniques that have already been developed. Furthermore, models for texture from other disciplines such as petrography have also been mentioned. This broadens the view of texture taken by researchers in computer vision.

- A scheme for creating compositions of primitive textures has been presented. This will allow the creation of complex textures from simple ones. Currently, no such compositional scheme for texture is available.

7.3 Future Work

Given the scope of this book, there are many potential benefits that could result from the application of these ideas. Here are a few suggestions arranged in increasing order of complexity.

7.3.1 APPLICATION TO INSPECTION PROBLEMS

In the short term, the algorithms in this book could be directly applied to practical inspection problems. Several examples have been already considered in this thesis, and potential areas for application have been identified.

As we have shown, humans are rather poor at describing texture. A more difficult problem is to encourage personnel in the manufacturing field to employ some of the taxonomy presented in this book. However, this requires a cultural change, and goes beyond the realm of engineering!

7.3.2 AN AUTOMATIC TEXTURE INTERPRETATION SCHEME

In chapter 6 we discussed compositional schemes to create arbitrarily complex textures. However, the exact computational scheme that can unravel the compositional structure of an arbitrary texture has not been implemented. We have mentioned a possible direction that could be explored to yield such a computational scheme. This would involve texture segmentation followed by an appropriate sequencing of the operators, as speculated in chapter 6.

7.3.3 TAXONOMIES FOR OTHER VISUAL CUES

It would be of interest to compile similar taxonomies for other visual cues also. For instance, work done by Green and Sun [41] is in the same spirit as this book as it tries to attach computational definitions to terms in the English language that represent motion.

Once the components of each visual cue have been analyzed in this manner, one could hope to create a visual language, as envisioned by Dondis [31].

7.3.4 Improving algorithms for Oriented Texture Analysis

Though the algorithms presented in this book appear to work well, there is room for further improvement. Here are some features that need more attention.

1. The sizes of the filters used to analyze the image are chosen *a priori*. It would be nice to estimate these filter sizes directly from the image.

2. Robust statistical techniques could be used for the least squares problem in chapter 3.

3. A region growing approach could be used to obtain a cleaner segmentation of the different flow images.

Appendix A

Rotation Invariance of D and q

We give a proof for the fact that D and q, as defined in section 3.7 are invariant with respect to rotation. Also, we determine the angle of rotation necessary to eliminate the term r. We use the notation presented in equation 3.29 for this analysis.

Consider the change of coordinates as defined in equation 3.10, of section 3.3.2. Since we are concerned with a rotation, the matrix $\mathbf{M}$ in this case is

$$\mathbf{M} = \begin{bmatrix} cos\theta & -sin\theta \\ sin\theta & cos\theta \end{bmatrix} \tag{A.1}$$

where θ is the angle of counter-clockwise rotation. Suppose the matrix $\mathbf{A}$ in the original coordinate system was

$$\mathbf{A} = \begin{bmatrix} a_{11} & a_{12} \\ a_{21} & a_{22} \end{bmatrix} \tag{A.2}$$

Then the matrix $\mathbf{A}'$ in the rotated coordinate system is given by

$$\mathbf{A}' = \mathbf{M}^{-1}\mathbf{A}\mathbf{M} = \begin{bmatrix} cos\theta & sin\theta \\ -sin\theta & cos\theta \end{bmatrix} \begin{bmatrix} a_{11} & a_{12} \\ a_{21} & a_{22} \end{bmatrix} \begin{bmatrix} cos\theta & -sin\theta \\ sin\theta & cos\theta \end{bmatrix} \tag{A.3}$$

or

$$\mathbf{A}' = \begin{bmatrix} a'_{11} & a'_{12} \\ a'_{21} & a'_{22} \end{bmatrix} \quad where$$

$$\begin{aligned} a'_{11} &= a_{11}cos^2\theta + a_{12}sin\theta cos\theta + a_{21}sin\theta cos\theta + a_{22}sin^2\theta \\ a'_{12} &= -a_{11}sin\theta cos\theta + a_{12}cos^2\theta - a_{21}sin^2\theta + a_{22}sin\theta cos\theta \\ a'_{21} &= -a_{11}sin\theta cos\theta - a_{12}sin^2\theta - a_{21}cos^2\theta + a_{22}sin\theta cos\theta \\ a'_{22} &= a_{11}sin^2\theta - a_{12}sin\theta cos\theta - a_{21}sin\theta cos\theta + a_{22}cos^2\theta \end{aligned} \tag{A.4}$$

From equation A.4 one can see that the quantities $D = a_{11} + a_{22} = a'_{11} + a'_{22}$ and $q = a_{21} - a_{12} = a'_{21} - a'_{12}$ are invariant with respect to rotation. The quantity r in the new coordinate system is

$$r' = a'_{21} + a'_{12} = (a_{22} - a_{11})sin2\theta + (a_{12} + a_{21})cos2\theta \tag{A.5}$$

Setting $r' = 0$ we obtain

$$tan2(\theta) = \frac{(a_{12} + a_{21})}{(a_{11} - a_{22})} \tag{A.6}$$

This equation determines the angle of rotation to make the term r' zero. This shows that for any matrix $\mathbf{A}$, one can always rotate the axes to eliminate the term r.

Appendix B

Region Refinement

The initially segmented image may look cluttered due to the presence of one-pixel sized holes, protrusions and so on. Such one-pixel regions are the result of noise effects and can be ignored. In order to perform this operation, we modified the 3x3 window region refinement operations presented in [13, p. 220] in order to handle images that are not binary.

The 3x3 window region refinement operation performs the following: (1) fills holes that are one pixel in size, based on the surrounding value (2) fills pixel-size notches in regions. Figure B.1 shows the changes made to specific types of regions. This operation is implemented by examining the 8 pixels surrounding a central pixel, and detecting certain pre-defined configurations of these pixels, as in figure B.1.

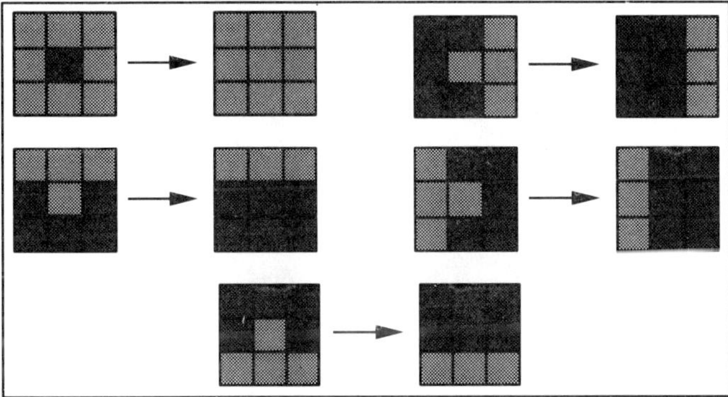

FIGURE B.1. Some cases of 3x3 window region refinement operations

Appendix C

Preparation of the manuscript

This entire book was processed in Latex, printed in Postscript, and did not involve any cut and paste operations.

Some of the more interesting images are the overlays of the orientation field on the original image (chapters 2 and 3). These were produced by drawing line segments on the image using Postscript.

Graphs were produced using the package GRAFIC, an interactive graphics program, written by John Dannenhoffer, and obtained from MIT. Three dimensional plots were produced using the package VIEWX, written by K. P. Beier of the University of Michigan. Charts were drawn on a Macintosh, and converted to Postscript.

Bibliography

[1] N. Ahuja and A. Rosenfeld, "Mosaic models for textures," *IEEE Trans. Pattern Analysis and Machine Intelligence*, vol. 3-(1), pp. 1–11, January 1981.

[2] J. Aloimonos and M. J. Swain, "Shape from texture," in *Int. Joint. Conf. Artificial Intelligence*, pp. 926–931, August 1985.

[3] J. Aloimonos, "Shape from texture," *Biological Cybernetics*, vol. 58, no. 5, p. 345, 1988.

[4] J. Aloimonos and M. Swain, "Shape from patterns: Regularization," *Int. J. Computer Vision*, vol. 2, pp. 171–187, 1988.

[5] A. A. Andronov *et al Qualitative theory of second-order dynamic systems*, John Wiley, New York, 1966

[6] V. I. Arnold. *Ordinary Differential Equations.* MIT Press, Cambridge, Massachusetts, 1973.

[7] D. K. Arrowsmith and C. M. Place. *Ordinary Differential Equations: A qualitative approach with applications.* Chapman and Hall, New York, 1982.

[8] H. Barrow and J. Tenenbaum, "Recovering intrinsic scene characteristics from images," in *Computer Vision systems*, pp. 3–27, Academic Press, 1978. (A. Hanson and E. Riseman eds).

[9] G. K. Batchelor. *An introduction to fluid dynamics.* Cambridge University Press, New York, 1967.

[10] J. Beck, K. Prazdny, and A. Rosenfeld. *A Theory of Textural Segmentation.* Academic Press, 1983.

[11] J. Beck, "Textural segmentation, second-order statistics, and textural elements," *Biological Cybernetics*, vol. 48, pp. 125–130, 1983.

[12] J. Beck, K. Pradzny, and A. Rosenfeld, "A theory of textural segmentation," in *Human and Machine Vision* (J. Beck, B. Hope, and e. A. Rosenfeld, eds.), pp. p1–38, Academic Press, New York, 1983.

[13] P. J. Besl. *Surfaces in early range image understanding*. PhD thesis, EECS Dept. University of Michigan, May 1986.

[14] P. J. Besl and R. C. Jain, "Segmentation through variable-order surface fitting," *IEEE Trans. Pattern Analysis and Machine Intelligence*, vol. 10, pp. 167–192, March 1988.

[15] P. J. Besl and R. C. Jain, "Three-dimensional object recognition," *Computing Surveys*, vol. 17, pp. 75–145, March 1985.

[16] D. Blostein and N. Ahuja, "Representation and three-dimensional interpretation of image texture: An integrated approach," in *First International Conference on Computer Vision*, pp. 444–450, 1987.

[17] A. C. Bovik, M. Clark, and W. S. Geisler, "Multichannel texture analysis using localized spatial filters," *IEEE Trans. Pattern Analysis and Machine Intelligence*, vol. 12, pp. 55–73, January 1990.

[18] J. M. Brady and H. Asada, "Smoothed local symmetries and their implementation," *Int. J. Robotics Research*, vol. 3, pp. 36–61, 1984.

[19] P. Brodatz, *Textures: A Photographic Album for Artists and Designers*. New York: Dover Publications, 1966.

[20] P. Brodatz, *Wood and Wood Grains: A Photographic Album for Artists and Designers*. New York: Dover Publications, 1966.

[21] J. F. Canny, "A computational approach to edge detection," *IEEE Trans. Pattern Analysis and Machine Intelligence*, vol. 8, pp. 679–698, November 1986.

[22] E. N. Coleman and R. Jain, "Obtaining 3-dimensional shape of textures and specular surfaces using four-source photometry," *Comp. Graph. and Image Processing*, vol. 18, pp. 309–328, 1982.

[23] R. Conners, C. W. McMillin, K. Lin, and R. E. Vasquez-Espinosa, "Identifying and locating surface defects in wood: Part of an automated lumber processing system," *IEEE Trans. Pattern Analysis and Machine Intelligence*, vol. 5, pp. 573–582, November 1983.

[24] R. W. Conners and C. A. Harlow, "Toward a structural textural analyzer based on statistical methods," *Computer Graphics and Image Processing*, vol. 12, pp. 224–256, 1980.

[25] H. Core, W. Cote, and A. Day, *Wood structure and identification*. Syracuse University Press, Syracuse N.Y., 1979.

[26] W. J. A. Dahm and K. A. Buch, "High resolution measurement three dimensional (256^3) spatio-temporal measurements of the conserved scalar field in turbulent shear-flows," Tech. Rep. GRI-5087-260-1443-4, Department of Aerospace Engineering, University of Michigan Ann Arbor, 1987. Presented at the Seventh Symposium on Turbulent Shear Flows, August 1989, Stanford, CA.

[27] L. S. Davis, S. Johns, and J. K. Aggarwal, "Texture analysis using generalized cooccurrence matrices," *IEEE Trans. Pattern Analysis and Machine Intelligence*, pp. 251–259, 1979.

[28] L. S. Davis, "Computing the spatial structure of cellular textures," *Comp. Graph. and Image Proc.*, vol. 11, pp. 111–122, 1979.

[29] J. G. Daugman, "Pattern and motion vision without laplacian zero crossings," *J. Opt. Soc. Am. A*, vol. 5, p. 1142=1148, July 1988.

[30] B. Dom, V. Brecher, R. Bonner, J. Batchelder, and R. Jaffe, "The P300: a system for automatic patterned wafer inspection," *Machine Vision and Applications*, vol. 1, pp. 205–221, 1988.

[31] D. A. Dondis, *A primer of visual literacy*. MIT Press, Cambridge, Massachusetts, 1973.

[32] D. J. Elliott, *Integrated Circuit Mask Technology*. New York: McGraw Hill, 1985.

[33] W. Erteld, H. Mette, and W. Achterberg, *Defects in Wood*. Leonard Hill, London, 1964.

[34] H. Farhoosh and G. Schrack, "Cns hls mapping using fuzzy sets," *Computer Graphics and Applications*, vol. 6, No. 6, pp. 28–35, June 1986.

[35] W. Fong, T. Cline, M. Walker, and S. Rosenberg, "A photolithography advisor: an expert system," in *Proc. SEMICON/EAST 85*, pp. 1–8, 1985.

[36] J. J. Gagnepain and C. Roques-Carmes, "Fractal approach to two-dimensional and three-dimensional surface roughness," *Wear*, vol. 109), pp. 119–126, 1986.

[37] B. S. Garbow, K. E. Hillstrom, and J. J. More, "Documentation for minpack subroutine hybrd1." Argonne National Laboratory, March 1980.

[38] L. Glass, Moire effect from Random dots. *Nature*, Vol. 223 pp. 578–580, 1969.

[39] L. Glass and R. Perez. Perception of random dot interference patterns. *Nature*, Vol. 246 pp. 360–362, 1973.

[40] P. A. Godinez, "Inspection of surface flaws and textures," *Sensors*, pp. 27–32, June 1987.

[41] M. Green and H. Sun, "A language and system for procedural modeling and motion," *Computer Graphics and Applications*, 1988.

[42] B. Grunbaum and G. C. Shephard, *Tilings and Patterns*. W. H. Freeman and Company, New York, 1987.

[43] L. G. C. Hamey, *Computer Perception of Repetitive Textures*. PhD thesis, Computer Science Department, Carnegie Mellon University, Pittsburgh, 1988.

[44] A. Hanson and E. Riseman, eds., *Computer Vision systems*. Academic Press, 1978.

[45] R. M. Haralick, K. Shanmugam, and I. Dinstein, "Textural features for image classification," *IEEE Trans. Systems, Man and Cybernetics*, vol. SMC-3 No.6, pp. pp610–621, November 1973.

[46] R. M. Haralick, "Statistical and structural approaches to texture," *Proceedings of the IEEE*, vol. 67, pp. 786–804, May 1979.

[47] N. Harris, *Form and Texture*. New York: Van Nostrand Reinhold Company, 1973.

[48] D. Hearn and M. P. Baker, *Computer Graphics*. Prentice Hall, 1986.

[49] D. J. Heeger, "Optical flow using spatiotemporal filters," *Int. J. Computer Vision*, vol. 1, pp. 279–302, 1988.

[50] J. Helman and L. Hesselink, "Representation and display of vector field topology in fluid flow data sets," *Computer*, pp. 27–36, August 1989.

[51] M. W. Hirsch and S. Smale, *Differential equations, dynamical systems, and linear algebra*. Academic Press, New York, 1974.

[52] B. K. P. Horn, *Robot Vision*. New York: McGraw-Hill, 1986.

[53] G. F. Horn, *Texture: A design element*. Worcester, Massachusetts: Davis Publications, 1974.

[54] I. Huntley and R. M. Johnson, *Linear and nonlinear differential equations*. Halsted Press, New York, 1979.

[55] K. Imaichi and K. Ohmi, "Quantitative flow analysis aided by image processing of flow visualization photographs," in *Proceedings of the Third International Symposium on Flow Visualization*, pp. 365–369, 1983.

[56] K. Imaichi and K. Ohmi, "Numerical processing of flow visualization pictures – measurement of two-dimensional vortex flow," *Journal of Fluid Mechanics*, vol. 129, pp. 283–311, 1983.

[57] I. Iwasaki and H. Tanaka, "On the flow visualization and turbulent measurement on the ripple models," in *Proceedings of the Third International Symposium on Flow Visualization*, pp. 681–685, 1983.

[58] A. K. Jain and R. C. Dubes, *Algorithms for Clustering Data*. Englewood Cliffs: Prentice Hall, 1988.

[59] R. Jain, A. R. Rao, A. Kayaalp, and C. Cole, "Machine vision for semiconductor wafer inspection," in *Machine Vision for Inspection and Measurement*, Academic Press, New York, 1989. ed. H. Freeman.

[60] R. Jain, T. Sripradisvarakul, and N. O'Brien, "Symbolic surface descriptors for 3-dimensional object recognition," in *Proceedings of SPIE*, January 1987.

[61] B. Julesz and J. R. Bergen, "Textons: The fundamental elements in preattentive vision and perception of textures," *Bell System Technical Journal*, vol. 62(6), pp. 1619–1645, 1983.

[62] K. ichi Kanatani, "Detection of surface orientation and motion from texture by a stereological technique," *Artificial Intelligence*, vol. 23, pp. 213–237, 1984.

[63] K. ichi Kanatani, "Structure and motion from optical flow under perspective projection," *Computer Vision, Graphics and Image Processing*, vol. 38, pp. 122–146, May 1987.

[64] J. N. Kapur, *Transformation Geometry*. Associated East-West Press (New Delhi), 1976.

[65] M. Kass and A. Witkin, "Analyzing oriented patterns," *Computer Vision, Graphics and Image Processing*, vol. 37, pp. 362–385, 1987.

[66] A. Kayaalp, A. R. Rao, and R. Jain, "Scanning electron microscope based stereo analysis," in *IEEE Conference on Computer Vision and Pattern Recognition*, (San Diego), June 1989.

[67] J. R. Kender, *Shape from Texture*. Marshfield, Mass.: Pitman Publishing.

[68] N. E. Kochin, I. A. Kibel, and N. V. Roze, *Theoretical Hydromechanics*. Interscience Publishers, New York, 1964.

[69] J. J. Koenderink and A. J. van Doorn, "Invariant properties of the motion parallax field due to the movement of rigid bodies relative to an observer," *Optica Acta*, vol. 22, pp. 773–791, 1975.

[70] F. Kollmann and W. Cote, *Principles of wood science and technology*. Springer-Verlag, New York, 1968.

[71] W. C. Leslie, *The Physical Metallurgy of Steels*. Hemisphere, 1981.

[72] J.G. Lee and W. G. Wee, "Detecting the spatial structure of natural textures based on shape analysis," *Computer Vision, Graphics and Image Processing*, vol. 31, pp. 67–88, 1985.

[73] M. Levine, *Vision in man and machine*. McGraw Hill, 1985.

[74] Jean Leymarie. *Qui etait, Van Gogh?* Editions d'Art Albert Skira, Geneva, 1968. in French.

[75] E. Lockwood and R. H. Macmillan, *Geometric Symmetry*. Cambridge University Press, 1978.

[76] S. Lefschetz, *Differential equations: geometric theory*. Interscience Publishers, New York, 1963.

[77] S. Y. Lu and K. S. Fu, "A syntactic approach to texture analysis," *Comp. Graph. and Image Proc.*, vol. 7, pp. 303–330, 1978.

[78] W. S. MacKenzie, *Atlas of Igneous Rocks and Their Textures*. John Wiley, 1982.

[79] J. Malik and P. Perona, "A computational model of texture segmentation," in *Conf. Proc. Computer Vision and Pattern Recognition*, pp. 326–332, 1989.

[80] B. B. Mandelbrot, *The fractal geometry of nature*. San Francisco, California: Freeman, 1983.

[81] B. B. Mandelbrot, *Fractals: Form, Chance and Dimension*. San Francisco, California: Freeman, 1977.

[82] K. V. Mardia, *Statistics of Directional Data*. Academic Press, New York, 1972.

[83] D. Marr, "Early processing of visual information," *Proc. Trans. Royal Phil. Soc., London, B*, vol. 275, pp. 483–524, 1976.

[84] D. Marr, *Vision, a computational investigation into the human representation and processing of visual information.* W.H. Freeman, New York, 1982.

[85] D. Marr and E. C. Hildreth, "Theory of edge detection," *Proc. R. Soc. Lond. B*, vol. 207, pp. 187–217, 1980.

[86] G. E. Martin, *Transformation Geometry.* Springer Verlag (New York), 1982.

[87] R. Mayer, *Dictionary of art terms and techniques.* Thomas Crowell Company, 1975.

[88] E. S. Meieran, P. A. Flinee, and J. R. Carruthers, "Analysis technology for VLSI fabrication," *Proceedings of the IEEE*, vol. 75, pp. 908–955, July 1987.

[89] W. Merzkirch, *Flow Visualization.* Academic Press, New York, 1974.

[90] R. Mohan and R. Nevatia, "Segmentation and description based on perceptual organization," in *Conf. Proc. Computer Vision and Pattern Recognition*, pp. 333–341, 1989.

[91] J. J. More, "The levenberg-marquardt algorithm, implementation and theory," in *Lecture Notes in Mathematics 630* (G. A. Watson, ed.), Springer-Verlag, 1977.

[92] J. Neter and W. Wasserman, *Applied linear statistical models.* Homewood, Illinois: Richard D.Irwin Inc., 1974.

[93] K. Noto, H. Ishida, and R. Matsumoto, "A breakdown of the karman vortex street due to the natural convection," *Proceedings of the Third International Symposium on Flow Visualization*, pp. 120–124, 1983.

[94] J. M. Ottino, *The kinematics of mixing: stretching, chaos, and transport.* New York: Cambridge University Press, 1989.

[95] J. Palis, Jr. and W. de Melo, *Geometric theory of dynamical systems,* Springer-Verlag, New York, 1982.

[96] A. Panshin, C. D. Zeeuw, and H. Brown, *Textbook of wood technology.* McGraw Hill, New York, 1964.

[97] A. Pentland, "Fractal-based description of natural scenes," *IEEE Trans. Pattern Analysis and Machine Intelligence*, vol. 6, pp. 661–674, November 1984.

[98] S. Petterssen, *Weather analysis and forecasting Vol 1: Motion and Motion Systems.* McGraw Hill, New York, 1956.

[99] K. Prazdny, "On the information in optical flows," *Computer Vision, Graphics and Image Processing*, vol. 22, pp. 239–259, 1983.

[100] W. H. Press, B. P. Flannery, S. A. Teukolsky and W. T. Vetterling, *Numerical Recipes in C: The Art of Scientific Computing*, Cambridge University Press, 1988.

[101] X. Qinghan and B. Zhaoqi, "An approach to fingerprint identification by using the attributes of feature lines of fingerprints," in *Eighth International Conference on Pattern Recognition*, pp. 663–5, October 1986.

[102] A. R. Rao and R. Jain, "Knowledge representation and control in computer vision systems," *IEEE Expert*, vol. Spring, pp. 64–79, 1988.

[103] A. R. Rao and R. Jain, "A classification scheme for visual defects arising in semiconductor wafer inspection," in *Third International Conference on Defect Recognition and Image Processing for Semiconductors*, (Tokyo), Sept 1989.

[104] A. R. Rao and R. Jain, "Knowledge representation and control in computer vision systems," *IEEE Expert*, vol. Spring, pp. 64–79, 1988.

[105] A. R. Rao and R. Jain, "Quantitative measures for surface texture description in semiconductor wafer inspection", *SPIE Conf. Microlithography*, March 1990, San Jose.

[106] A. R. Rao and B. G. Schunck, "Computing oriented texture fields," in *Proc. Conf. Computer Vision and Pattern Recognition*, 1989.

[107] T. Sallam and A. Mitsunaga, "Visualization and computational studies on the formation and stability of karman vortex streets," *Proceedings of the Third International Symposium on Flow Visualization*, pp. 125–129, 1983.

[108] B. G. Schunck, "Gaussian filters and edge detection," *Research Publication GMR-5586, Computer Science Department, General Motors Research Laboratories*, October 1986.

[109] B. G. Schunck, "Edge detection with Gaussian filters at multiple scales," Proc. Workshop on Computer Vision, 1987.

[110] A. Spry, *Metamorphic Textures*. Pergamon Press, 1969.

[111] K. A. Stevens, "Computation of locally parallel structure", Biological Cybernetics, Vol. 29, pp.19-28, 1978.

[112] S. Strassmann, "Hairy brushes," *Computer Graphics*, vol. 20, no. 4, pp. 225–232, 1986.

[113] R. Szymani and K. McDonald, "Defect detection in lumber: state of the art," *Forest Products Journal*, vol. 31(11), pp. 24–44, November 1981.

[114] H. Tamura, S. Mori, and T. Yamawaki, "Textural features corresponding to visual perception," *IEEE Trans. Systems, Man, and Cybernetics*, vol. 8, pp. 460–473, June 1978.

[115] F. Tomita, Y. Shirai, and S. Tsuji, "Description of textures by a structural analysis," in *Int. Joint. Conf. Artificial Intelligence*, pp. 884–889, 1979.

[116] F. Tomita, Y. Shirai, and S. Tsuji, "Description of textures by a structural analysis," *IEEE Trans. Pattern Analysis and Machine Intelligence*, vol. 4-(2), pp. 183–191, 1982.

[117] G. T. Toussaint, "The Relative Neighbourhood Graph of a Finite Planar Set," Pattern Recognition, Vol 12, pp. 261-268, 1980.

[118] M. Van Dyke, *An Album of Fluid Motion*. Parabolic Press, 1982.

[119] A. Verri and T. Poggio, "Motion field and optical flow: qualitative properties," *IEEE Transactions on Pattern Analysis and Machine Intelligence*, vol. 11, pp. 490–498, 1989.

[120] F. M. Vilnrotter, R. Nevatia, and K. E. Price, "Structural analysis of natural textures," *IEEE Trans. Pattern Analysis and Machine Intelligence*, vol. 8, pp. 76–89, January 1986.

[121] H. Voorhees, "Finding texture boundaries in images," *MIT AI lab, Technical Report 968*, June 1987.

[122] H. Voorhees and T. Poggio, "Detecting textons and texture boundaries in natural images," in *First International Conference on Computer Vision*, pp. 250–258, 1987.

[123] V. Voort, *Applied Metallography*. New York: Van Nostrand Reinhold, 1986.

[124] J. Warnock, ed., *POSTSCRIPT Language reference manual*. Addison Wesley, 1987. Adobe Systems Inc.

[125] H. Wechsler, "Texture analysis: A survey," *Signal Processing*, vol. 2, pp. 271–282, 1980.

[126] H. Weyl, *Symmetry*. Princeton, 1952.

[127] A. P. Witkin, "Recovering surface shape and orientation from texture," *Artificial Intelligence*, vol. 17, pp. 17–45, 1981.

[128] G. Wyszecki and W. S. Stiles, *Color Science.* New York: Wiley, 1967.

[129] W. Yang, "Preface," in *Proceedings of the Third International Symposium on Flow Visualization*, pp. 681–685, 1983.

[130] K. M. Yip, "KAM: Automatic Planning and Interpretation of Numerical Experiments Using Geometric Methods," PhD Thesis, MIT Artificial Intelligence Laboratory, 1989.

[131] N. Yokoya, K. Yamamoto, and N. Funakubo, "Fractal-based analysis and interpolation of 3d natural surface shapes and their application to terrain modeling," *CVGIP*, vol. 43, No. 3, pp. 284–302, June 1989.

[132] W. H. Zachariasen, *Theory of X-Ray Diffractions in Crystals.* John Wiley, New York, 1945.

[133] C. T. Zahn, "Graph theoretical methods for detecting and describing Gestalt clusters," *IEEE. Trans. Computers*, Vol. C-20, No. 1, Jan. 1971.

[134] S. W. Zucker, "Toward a model of texture," *Comp. Graph. and Image Proc.*, vol. 5, pp. 190–202, 1976.

[135] S. W. Zucker and P. Cavanaugh, "Subjective figures and texture perception," Tech. Rep. TR-85-2R, Computer Vision and Robotics Laboratory, Department of Electrical Engineering, McGill University, 1985.

[136] S. W. Zucker, "Early Orientation Selection: Tangent Fields and the Dimensionality of Their Support", *Comp. Vision, Graphics and Image Proc.*, Vol 32(1), pp.74-103, Oct 1985.

[137] "Fractographic features revealed by light microscopy," in *Metals Handbook, 8th ed. Vol. 9: Fractography and Atlas of Fractographs*, Metals Park, Ohio: American Society for Metals.

[138] "Jeol news," 1986. Vol 24E, No. 3.

[139] Surface texture (surface roughness, waviness, and lay). *The American National Standards Institute*, 1985. ANSI/ASME B46.1.

Index

Permissions

I wish to credit the following sources for permission to reproduce figures.

Figure 1.5 and figure 3.24 from *Integrated Circuit Mask Technology* by D. J. Elliott, pg. 254, ©1985 McGraw Hill.

Figures 1.6, 2.13, 2.8, 2.9, 5.3 and 6.8 from *Textures: A Photographic Album for Artists and Designers* by Phil Brodatz, 1966, ©Lillian Brodatz.

Figures 2.4, 2.5, 3.1 and 3.16 from *An Album of Fluid Motion*, Figure 31, ©1982 Milton Van Dyke, Stanford University.

Figures 2.6, 2.7 and 3.13 ©D. H. Peregrine, Bristol University, (appeared in *An Album of Fluid Motion*, by Milton Van Dyke, Parabolic Press, 1982, as Figure 1 and Figure 3).

Figures 2.15 and 3.18 from *Defects in Wood*, by W. Erteld, W. H. Metta and W. Acterberg, 1964 (Leonard Hill) ©Blackie and Son Ltd., Glasgow and London.

Figure 2.16 from the article *"Identifying and locating surface defects in wood: part of an automated lumber processing system"*, by R. Conners *et al*, Figure 2 (j), pg. 575, ©1983 IEEE.

Figure 3.22 by F. N. M. Brown, courtesy of University of Notre Dame (appeared as Figure 75 in *An Album of Fluid Motion*, by Milton Van Dyke, Parabolic Press, 1982).

Figure 3.26: *A Starry Night*, 1889 by van Gogh, (in Kunsthalle, Bremen), Indian ink, 18 1/2 x 24 1/2 ″ (detail) from the book "Qui était van Gogh?" by Jean Leymarie; Geneva; ©Editions d'Art Albert Skira, 1968.

Figure 5.3 from *Metals Handbook*, Vol 9, 8th Ed., *Fractography and Atlas of Fractographs*, American Society for Metals, Metals Park, OH, 1974.

Viscous Fluid
flow
White?